THE RUNAWAY UNIVERSE

OTHER BOOKS BY DONALD GOLDSMITH

Voyage to the Milky Way: The Future of Space Exploration

The Ultimate Einstein

Worlds Unnumbered: The Search for Extrasolar Planets

The Hunt for Life on Mars

Einstein's Greatest Blunder? The Cosmological Constant and Other Fudge Factors in the Physics of the Universe

The Search for Life in the Universe
(with Tobias Owen)

The Astronomers

Mysteries of the Milky Way
(with Nathan Cohen)

Space Telescope: Eyes Above the Atmosphere
(with George Field)

Nemesis: The Death-Star and Other Theories of Mass Extinction

THE
RUNAWAY
UNIVERSE

*The Race to Find
the Future of the Cosmos*

DONALD GOLDSMITH

PERSEUS PUBLISHING
Cambridge, Massachusetts

Many of the designations used by manufacturers and sellers to distinguish their products are claimed as trademarks. Where those designations appear in this book and Perseus Publishing was aware of a trademark claim, the designations have been printed in initial capital letters.

CIP information for this book is available from the Library of Congress.
ISBN: 0-7382-0429-3

Perseus Publishing is a member of the Perseus Books Group

Text design by Jeff Williams
Set in 11-point Palatino by the Perseus Books Group

1 2 3 4 5 6 7 8 9 10—03 02 01
First paperback printing, December 2000

Perseus Publishing books are available at special discounts for bulk purchases in the U.S. by corporations, institutions, and other organizations. For more information, please contact the Special Markets Department at HarperCollins Publishers, 10 East 53rd Street, New York, NY 10022, or call 1-212-207-7528.

Find us on the World Wide Web at http://www.perseuspublishing.com

To Rachel
and
to the cosmologists who make
the universe go 'round

CONTENTS

Acknowledgments   ix

1   The Runaway Universe   1

2   Einstein's Dilemma   7

3   The Discovery of the Galaxies   15

4   The Expansion of the Universe   25

5   The Inflationary Cosmos   49

6   Dark Matter Rules   61

7   Supernovae Reveal the Accelerating Universe   77

8   Why Stars Explode   97

9   The Race to Find the Future of the Universe   113

10   Could the Cosmological News Be Wrong?   141

11   The Cosmic Background Radiation   163

12   The Birth of Galaxies Reveals the Density of Matter   181

13   Gravitational Lenses Bend the Cosmos   191

14   How Can We Explain the Cosmological Constant?   201

15   Prospects for Resolving the Cosmic Mysteries   211

Further Reading   223

Index   225

ACKNOWLEDGMENTS

A host of astronomers have helped me in writing this book, as have friends and relatives who have stood firmly by me in this effort. The happy result of their kindness is a long list of people to whom I wish to express my gratitude. I thank Fred Adams, Jenni Adams, Anthony Aguirre, Halton Arp, John Bahcall, Neta Bahcall, Hans Bethe, Roger Blandford, Adam Burrows, Butler Burton, Per Carlson, John Carlstrom, Catherine Cesarsky, Diego Cesarsky, Marc Davis, Daniel Eisenstein, John Ellis, Richard Epstein, Gus Evrard, Wendy Freedman, Peter Garnavich, Margaret Geller, Dan Gezari, Gerson Goldhaber, Paul Goldsmith, Ariel Goobar, Leonid Grishchuk, Alan Guth, Mario Hamuy, Leif Hansen, Craig Hogan, Wayne Hu, John Huchra, Henning Jørgensen, Robert Kirshner, Chris Kochanek, Rocky Kolb, Lawrence Krauss, Andrei Linde, Mario Livio, Piero Madau, Steve Maran, Laurence Marschall, Giuseppe Monzo, Ken Nomoto, Henry Nørgaard, Hans-Ulrik Nørgaard-Nielsen, Alain Omont, Toby Owen, Franco Pacini, Jim Peebles, Carl Pennypacker, Martin Rees, Hubert Reeves, Pilar Ruiz-Lapuente, Bernard Sadoulet, Anneila Sargent, Wallace Sargent, Brian Schmidt, Frank Shu, Joe Silk, David Spergel, Hyron Spinrad, Paul Steinhardt, Nick Suntzeff, Max Tegmark, Tony Tyson, David Tytler, Marie-Hélène Ulrich, Robert Wagoner, J. Craig Wheeler, Stan Woosley, and especially Ken Brecher, Sean Carroll, George Field, Alex Filippenko, Saul Perlmutter, Adam Riess, and Michael Turner for their kind and courteous assistance, often at inconvenient times and with respect to matters they had explained before. My thanks also go to Kathy and Bruce Armbruster, Sam Bader, Aviva Brecher, Arlene and Pascal Debergue,

Susan Field, Carol and Richard Gammon, Marjorie and Victor Garlin, Jane Goldsmith, Rachel Goldsmith, Amy and George Gorman, Dolly and Don Hatch, Marjorie and Jerry Heymann, Annemarie and Hagen Kleinert, Sally Maran, Ellen Marschall, Lynn and Jean-Pierre Merle, Martha and Ken Mostow, Christine and Craig Nova, Eleanor and Sandy Orr, Merrinell Phillips, Sheryl Reiss, Brian Salzberg, and Sara and Roy Schotland for their encouragement and support. At Perseus Books, Amanda Cook, Marco Pavia, Jeff Robbins, and Lissa Warren have been a pleasure to work with.

THE RUNAWAY UNIVERSE

CHAPTER ONE

THE RUNAWAY UNIVERSE

IMAGINE A STRANGE UNIVERSE in which the expansion of the cosmos, instead of being slowed by gravity, undergoes a continuous acceleration from the presence of a mysterious form of energy. This energy, concealed from any direct detection by its complete transparency, permeates seemingly empty space, furnishing the cosmos with a "free lunch" of just the sort that old wives' tales forbid. Just as amazingly, every cubic centimeter of the new space that the ongoing cosmic expansion creates likewise teems with this invisible energy, the existence of which endows each volume of space with a tendency to expand. As a result, the universe multiplies its energy content many times over as time goes by. The increase in its hidden energy makes the universe accelerate ever more rapidly, eventually driving its basic units of matter to utterly unfathomable separations. Instead of a chance to contract, perhaps to recycle itself through another big bang, this universe faces a future in which all cosmic distances grow to billions of times their present immense values. As this happens, the average density of matter in the cosmos falls ever more rapidly toward zero, because the energy of empty space makes the universe expand at a continuously increasing rate.

This parallel universe is our own—if astronomers have correctly interpreted their recent observations. They have known for seventy years that the universe is expanding: Clusters of galaxies, each a giant agglomeration of matter containing billions upon billions of stars, are moving away from one another throughout all of space. Indeed, space itself must be expanding, carrying the galaxy

1

clusters with it. This cosmic expansion implies that the universe began, at least in its present phase, at a time when all the matter (and all space, too!) existed at a moment of near-infinite temperature and density, which astronomers call the "big bang." Ever since the big bang, now estimated to have occurred about 14 billion years ago, the universe has cooled while expanding. This cooling has allowed some of the matter, which at first spread smoothly throughout space, to agglomerate through gravitation into the clumps that became galaxies, stars, and the relatively tiny objects we call "planets" and "moons."

THE RUNAWAY UNIVERSE

Since 1929, when Edwin Hubble discovered the expansion of the universe, astronomers have confronted the burning issue of whether this expansion will continue forever or whether the expansion may someday reverse itself into a contraction that might lead to another big bang. For more than eighty years, since Albert Einstein first created what still appears to be the correct theory of how space in the universe behaves, we have known that the amount of matter in the universe determines its future. Only one phenomenon might someday reverse the expansion to produce a universal contraction: gravity. Astronomers know that the gravitational attraction among all the objects in the universe has already slowed its expansion. The crucial question of whether this attraction will someday actually reverse the expansion, making the universe start to contract, finds its answer in the average density of matter. If that density, the amount of mass contained in a standard volume of space, exceeds a certain critical value, then the universe must eventually contract. If not, the expansion will continue indefinitely.

So astronomers believed until 1998. In that year, astronomers obtained startling new evidence from exploding stars (known by their Latin name as "supernovae"), seen in far-distant galaxies, that resurrected a long-discarded notion that Albert Einstein had created. In 1917, Einstein introduced an additional term, quickly named the "cosmological constant," into his equations describing the behavior of the cosmos. He did so for what seemed the best of reasons: the need to explain a "static universe," one in which space neither expands nor contracts. At that time, no astronomer suspected that a universal expansion might exist. Einstein, however, perceived that his equations, in the absence of a cosmological constant, imply that the cosmos must either expand or contract throughout cosmic history. His cosmological constant—not only permissible, but in a mathematical sense mandatory—allows the full spectrum of possibilities: expansion, contraction, or a static universe, with the latter permitted only if the constant has a single particular value. When Hubble's observations led to the discovery of the expanding universe, Einstein and his fellow scientists happily assigned the constant a value of zero, leaving it technically in existence but of no practical effect, and discarded the concept of a static universe.

For seven decades, this zero value seemed correct. Physicists who attempted to deduce the constant's value on theoretical grounds could do no better than to conclude that it ought either to be zero or to have values so enormously large that the universe could not exist. This analysis favored the zero value. Yet observations appear to have bypassed this reasoning. As astronomers improved their abilities to detect and to study supernovae that have exploded in galaxies billions of light-years beyond the Milky Way, they acquired the ability to discriminate between two crucial factors affecting the expansion: the amount of matter in the universe and the cosmological constant.

Whereas the gravitational forces among objects with mass act to slow the expansion, a cosmological constant greater than zero tends to make the cosmos expand more rapidly. This fact allows a cosmological constant with one particular value to balance the effect of gravity exactly—Einstein's original motivation for its introduction. Other nonzero values of the cosmological constant, however, imply a cosmos in which matter slows the expansion while the cosmological constant accelerates it, but the two effects do not balance each other exactly. By observing supernovae at immense distances, astronomers can hope to see the net effects of the contest between these two effects. A cosmological constant with a value greater than zero acts to increase the rate of expansion above what we would find if the constant equals zero. In that case, astronomers who observe supernovae in faraway galaxies should find that the exploding stars have greater distances from us than we would expect in a universe with a constant equal to zero. They can also measure the size of the cosmological constant, because a larger constant will have produced a greater acceleration of the expansion during the time since the supernovae exploded. This greater acceleration will have put still more "extra" distance between ourselves and the supernovae.

Early in 1998, two groups of supernova observers stood the cosmological world on its ear by announcing that their data analysis had in fact revealed a nonzero value for the cosmological constant. The implications for the future of the universe are tremendous—so significant, in fact, that in all that follows, we must bear in mind that the results from supernova observations must pass the test of skeptical scrutiny before we incorporate them in our inner fibers. If the cosmological constant has a nonzero value, the universe will expand forever, and, indeed, it will expand ever more rapidly as time goes by. Despite matter's heroic efforts to reduce the expansion rate to zero through gravitation, which have succeeded in slowing the expansion somewhat during the past 14 billion years, the cosmological constant's tendency to accelerate the cosmic expansion must eventually triumph.

The acceleration will win because the effects of gravity grow weaker as the universe expands, separating clusters of galaxies by greater distances and thus reducing their mutual gravitational attraction. In contrast, the cosmological constant keeps on coming: Every newly created cubic centimeter of space appears with the same amount of energy as all the cubic centimeters that already exist. A universe with a cosmological constant has the ability to produce new energy continuously, literally from nothing! The inevitable victory of acceleration over deceleration implies that in the long run, the universe will expand ever more rapidly. If the cosmological constant has a value greater than zero, then in epochs that lie tens and hundreds of billions of years in our future, the universe will expand far more rapidly than now—faster and still faster, so that the expansion will produce a "runaway," in which clusters of galaxies separate from one another at ever-greater velocities.

The runaway universe has produced relatively few enthusiasts in professional circles or among the general public. Aside from the astronomical difficulties of explaining why the cosmological constant should not equal zero, or why it should have the specific value implied by the recent observations of distant supernovae billions of light-years beyond the Milky Way, most of the public finds

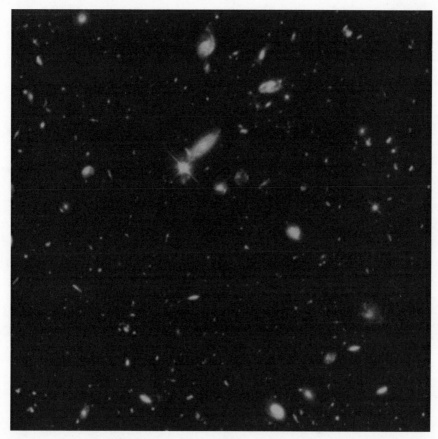

This photograph, the longest-exposure image ever taken by the Hubble Space Telescope, shows galaxies more than 10 billion light-years away. The light from these galaxies, which has taken billions of years to reach us, reveals conditions in the universe billions of years ago, when all galaxies were young. (Photograph courtesy of the Space Telescope Science Institute.)

a bit distasteful the notion that the expansion will not only continue, but will in fact proceed at an ever-greater rate. To this, the supernova observers—and scientists in general—have a reply: Try to get used to it. If we seek to uncover a truth independent of our individual biases and beliefs, neither we nor professional scientists can let our desires rule cosmology, the study of the universe as a whole. What we must do, and what scientists love to do, is to pay close attention to the latest observations and their interpretation, probing for different explanations of the data. Only when we have satisfied ourselves that a nonzero cosmological constant offers by far the most coherent way to interpret the observational facts should we embrace the concept of the runaway universe. Even then, we must remain aware that new data and new interpretations may soon appear, causing us once again to question the framework within which we conceive the cosmos.

Let us examine, then, the cosmological observations and theories that have brought astronomers to the concept of the runaway universe, along with attempts to explain the accelerating expansion and the prospects for future resolution of the key cosmic issues. We shall meet observations of distant supernovae and the two teams of astronomers who compete to find the secrets of the cosmos, as well as other ways to attack these mysteries, including the bending of light by gravity, the formation of galaxies billions of years ago, and the faint afterglow of creation, which carries information about the universe at a time only a few hundred thousand years after the big bang. The mind-bending concepts involved in this examination will provide excellent mental exercise, and the results will prove so amazing that your friends and family will doubt what you have to tell them. Yet that is the cosmos you are meeting—not a parallel universe, but apparently our own.

CHAPTER TWO

EINSTEIN'S DILEMMA

A LONG LIFETIME AGO, as generals lacking new ideas ordered soldiers from their trenches to near-certain death during the worst days of World War I, Albert Einstein sat in Berlin, pondering the universe. Appointment to the Prussian Academy of Sciences in 1914, a few months before the war began, had provided Einstein with a good salary, an office, and no teaching obligations. Rarely has governmental support for basic research yielded greater dividends. Einstein had left behind in Switzerland his first wife, Mileva, and their two sons, the younger one destined to a lifetime of schizophrenia, and had found happiness in Berlin with his second wife. Elsa, a few years older than he, had known him from childhood, since she was both his first and second cousin, and watched over her "Albertle" (little Albert) with tender care. Einstein needed it, for his fellow scholars in Germany tended to shun him as a pacifist Jewish outsider, adept in the realms of new physics but not a real German at all. Far from displaying the hearty patriotism that his colleagues expected and flaunted (at least during the early months of the war), Einstein had put himself beyond the pale as the war began: He had circulated, with zero success, an antiwar petition opposing the manifesto that almost every other prominent scientist in Germany had signed in support of Germany's invasion of Belgium.

THE GENERAL THEORY OF RELATIVITY

Einstein, who had renounced his German citizenship while still a teenager (only to receive it again automatically through his election to the Prussian Academy of Sciences) was accustomed to being regarded as an oddball. What counted most to him was the opportunity to work undisturbed. During the first half of the war, he had brought to completion his general theory of relativity, a new way to regard space and time in the universe, on which he had labored for five long years. Now, during the winter and spring of 1917, as Germany's submarine attacks on neutral shipping provoked the American declaration of war that would decide the world contest, Einstein struggled with the dilemma arising from the equations that he had conceived to describe the cosmos.

According to Einstein's general theory of relativity, what we call "gravity" can best be understood as the bending of space, distorted most of all in the immediate neighborhoods of objects with mass. Bent space tells matter how to move through it. The sun, for instance, bends space in its vicinity, so that the Earth and the sun's other planets tend to roll toward it. Each planet's momentum allows it to roll around and around the sun, like a fast-moving marble circling the side of a bowl. To Einstein, and to the generations of physicists that followed him, this description of space bent by gravity has a beauty of its own, a melding of concepts that explains what we already observe (planetary orbits and other responses to gravity) while providing new ways to imagine gravitational forces, plus new predictions of how gravity affects matter and radiation.

Einstein also saw that his general theory of relativity provides a description of the entire universe, the totality of all space and everything in it. But in examining the solutions for his equations, Einstein discerned an apparent impossibility, a potentially fatal flaw that cried out for correction so that his theory would correspond to the reality that astronomers observed. According to Einstein's equations as he first wrote them, the universe cannot exist in a static form. Instead, at any moment, all of space must be either expanding or contracting, and it could never remain at rest, the state that everyone in 1917 expected, since nothing astronomers had seen suggested a universal expansion or contraction.

So the greatest physicist of the modern era sat in his office, warmed against the dull, gray winter of Brandenburg, turning over the possibilities. In the privacy of his study, Einstein almost discovered the expansion of the universe, not by using the mighty telescope in California—just then nearing completion—that Edwin

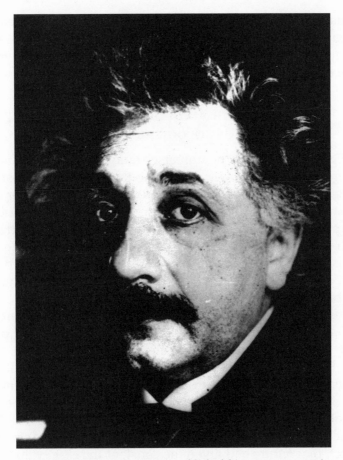

Albert Einstein (1879–1955) published his greatest contributions to physics, the special and general theories of relativity, in 1905 and 1916, respectively. (Photograph courtesy of the National Archives of the Neils Bohr Library of the American Institute of Physics.)

Hubble would employ a decade later to make this discovery, but through the power of his mind alone. But Einstein, though never afraid to overturn received opinion, left this door unopened. Influenced by the weight of astronomical observations, which had recorded no overall motions in the cosmos, he concluded that his equations must be incomplete. If the observational data did not agree with the general theory of relativity—more precisely, with the equations describing the universe that follow directly from that theory—then he must have left something out, some aspect of the equations that would leave space static and motionless. And so Einstein achieved what appeared to be a colossal insight into the nature of space itself.

THE BIRTH AND (FIRST) DEATH OF THE COSMOLOGICAL CONSTANT

Seeking a way to keep the universe static, Einstein perceived that his fundamental equations describing the space in the cosmos allow the existence of an additional constant term. Whenever mathematicians "integrate" an equation, smoothly summing an infinite series of terms, they create the possibility—indeed, the certainty— that their result will include a "constant of integration." To evaluate this constant, they must refer to other aspects of the problem they are solving. Quite often, the constant of integration turns out to be zero, which means that it has only a technical claim to existence. Einstein saw that in his crucial equation, the constant might well have a nonzero value: Nothing in the theories of physics known to him could suggest or rule out any values at all, though he could see that enormously large values would not allow the universe to exist in the form that we see it. Within the immense range of possible values that remained for the constant, one particular value would allow space to exist without either expanding or contracting. Einstein therefore published a scientific paper stating, in effect, that nature had given the constant the exact value that allows a static universe to exist. This conclusion reduces all other values for the constant to mere theoretical possibilities, conceivable but of no realistic importance. Soon after Einstein published his conclusions, the constant in his equation received the appropriately elevated title of the "cosmological constant."

Mathematically, Einstein's analysis seemed unquestionable, but lurking within the seemingly innocuous cosmological constant lay a new world of physics. Instead of discerning the expansion of the universe, Einstein had apparently discovered new properties of a most fundamental aspect of the cosmos: the nature of space itself. A nonzero, positive cosmological constant implies unambiguously that empty space contains a stunning surprise: Although it looks empty, and in fact is empty to every technique of measurement, every cubic centimeter of space abounds with invisible energy! This energy, utterly invisible and unperceivable, has no "useful" form.

Its sole function and result is to push space apart, and as it does so, new space comes into existence, itself just as rich in energy as the space that gave it birth. The cosmological constant therefore amounts to a "free lunch" of energy. As space expands, each new volume of space contains more of the energy that the constant provides. (Note to those who dream of tapping this energy: First solve the problem of perpetual motion, and then we'll talk.)

But if the universe has just the right balance, space will not expand. Instead, as Einstein showed (and this was, after all, the reason for his consideration of nonzero values for the cosmological constant), the appropriate value of the constant will allow the universe to balance its tendency to contract, as all parts of the cosmos attract one another gravitationally, with its tendency to expand. The cosmological constant cannot stop an ongoing contraction, but it can maintain an equilibrium between the universe's expansile and contractile propensities. Einstein knew, of course, that his cosmological constant amounted to no more than a hypothetical truth, pending further data that would confirm or deny its existence. In the paper that he submitted to the leading German physics journal in February 1917 (where it was published a week later, since the journal's editorial offices were located in Berlin and relatively little was happening in the world of physics), Einstein stated that "whether [the hypothesis of a cosmological constant] can be maintained from the standpoint of current astronomical knowledge will not be investigated here." In a letter to his old school friend Michelangelo Besso, he wrote that "I should have, in the spirit of Newton, set [the cosmological constant] equal to zero. But the new understanding speaks in favor of a nonzero [cosmological constant], which allows a nonzero density of matter to be introduced [without the universe contracting]."

In April, Einstein wrote to the Dutch cosmologist Willem de Sitter, replying to de Sitter's assertion that the cosmological constant did not make physical sense and that its value could not be measured. Einstein defended his work, stating that "the postulate of general relativity *requires* the introduction of the [cosmological constant] into the field equations. It will be our factual knowledge of the composition of the starry heavens, of the apparent motions of the stars, and of the state of spectral lines as a function of conditions far from us that will allow us empirically to answer the question whether the [cosmological constant] equals zero or not. Conviction is a good mainspring, but a bad judge!" Here Einstein was quite correct. It seems clear from his later discussions, however, that for several years Einstein believed in a nonzero cosmological constant, a hitherto-unknown aspect of space with the near-magical property of hiding energy, capable of explaining why the universe is neither expanding nor contracting.

How utterly wrong Einstein was, and yet how right—in a way—he nevertheless proved to be! The motivation for his introduction of the cosmological constant turned out to be fallacious, for space in the universe actually is expanding.[1] When this became clear, Einstein rued his creation, characterizing it to George Gamow, a younger colleague who gained fame as a physicist and a popularizer of science, as his "greatest blunder." The discovery of the expanding universe led Einstein, Gamow, and all other cosmologists to conclude that even though Einstein's key equation does allow for a nonzero cosmological constant, which in theory might have any value, the actual value of the constant should be zero: The observational results require nothing else, and no good reason for a nonzero constant exists in the world of physics.

During the next sixty years, as various apparent conflicts arose between cosmological theory and new observations, astronomers and cosmologists occasionally invoked the constant as a possible means, not much liked and often scorned by others, of explaining these apparent discrepancies. The suggested values for the constant differed from Einstein's original one, chosen to make the universe static; instead, these values tended to allow the expansion of the universe to slow down for a few billion years and then to increase its rate of expansion. When better observations resolved the conflicts by demonstrating that a universe with a zero cosmological constant could fit the observational data quite well, cosmologists sighed with relief. Through most of the twentieth century, Einstein's cosmological constant seemed to offer a constant temptation to those astronomers who failed either to perceive the bare beauty of Einstein's equation without the constant or to believe that improved observations would eventually validate the description of the cosmos in which the constant equals zero.

---

[1]As the Russian mathematician Alexander Friedmann soon demonstrated, the cosmological constant cannot really keep the universe in a static state, because the constant produces an unstable balance between the universe's tendency toward either expansion or contraction: The slightest deviation from a perfect balance would cause the universe to expand or contract forever.

THE NEW COSMOLOGICAL CONSTANT

Thus when new observations during the final years of the twentieth century once again called for a nonzero cosmological constant, the astronomical community displayed a predictable reaction. Wait a little while, said the wiser heads, and the constant will once again prove to be zero. But now, as the 9's roll into 0's, much of this reaction has played itself out. Unlike the earlier news of a nonzero constant, the cosmological constant now appears here to stay, with a nonzero value once again different from the number that Einstein had deduced on different grounds.

This new value has implications just as shocking as the one that Einstein suggested. As always in science, we would do well not to accept it as fully validated without new sets of observations that confirm today's conclusions. Because the existence of a nonzero cosmological constant calls for a revision in how we think about space, and because we lack good theoretical grounds for believing in a constant with the value that the observations now imply, scientists continue to follow their usual, conservative approach, which has served them well in revising their views of the universe. In the long run, observations will always triumph over theory, but in the short term, the reverse often proves true. Before astronomers will abandon a theory that has served them well and agrees with a wealth of data, they will demand more than a few striking new observations. If these new observations imply not only the modification of a key theory, but also that we live at a particular time in the universe's history—as the new observations do—then they merit double suspicion, because scientists understandably prefer to conclude that our moment in time, like our location in space, is a random one. The observations implying a nonzero cosmological constant have therefore faced an uphill battle for acceptance. The fact that this battle appears to be nearly won testifies to the inherent strength of the observational data.

To understand the challenge that the new results offer and the reasons for astronomers' suspicions, we must follow the key steps that led to the realization that we live in an expanding universe of galaxies and galaxy clusters. This news came to us from hard-won observational data, much of it obtained by a single individual, working at the world's finest astronomical observatory on a mountaintop in southern California.

CHAPTER THREE

THE DISCOVERY OF
THE GALAXIES

EDWIN HUBBLE KNEW LITTLE OF, and apparently cared less for, the world's theoretically oriented physicists and astronomers, who perform their research on paper rather than by examining the world around them. Born in Missouri, educated in Illinois, and appointed a Rhodes scholar shortly before World War I, Hubble was a tall, athletic Midwesterner, intrigued by science at the University of Chicago but sufficiently practical to study law—until he suddenly decided on science after all. After enlisting in the army in May 1917, soon after Einstein had published his conclusions about the cosmological constant (in February) and America had entered the war (in April), Hubble went overseas in September 1918, though he never tasted actual combat, a fact that his later interlocutors, to whom he often proclaimed the pain of having to go forward into battle without stopping to help the wounded, would have been hard-pressed to discover from his conversation. Back in the United States, Hubble resumed his astronomical research, which had won him a highly desirable appointment as a member of the permanent staff at the Mount Wilson Observatory, the site of the world's finest telescopes.

THE MOUNT WILSON OBSERVATORY AND
THE LIMITS OF THE MILKY WAY

Today, nearly blinded by city lights, the observatory on Mount Wilson overlooks the crowded Los Angeles basin, capable of solar observations but bereft of its former importance, which has passed to the mighty telescopes in Hawaii, Chile, and on other mountaintops with dark skies, clear air, and few clouds. Eighty years ago, however, with Hollywood not yet the center of the movie industry, Mount Wilson seemed well isolated from the small city below it. Mule trains toiling up the mountain's slope brought most of the observatory's supplies, as well as the construction materials for the 60-inch and 100-inch reflecting telescopes. One of the mule drivers, a teenager named Milton Humason, demonstrated an aptitude in tinkering with the observatory's equipment that led to his being hired as a general assistant. Hubble liked Humason, who knew his place in the hierarchy and rarely failed to address the astronomer with his former military title of "Major." Eventually, Humason would take most of the photographic plates that confirmed Hubble's conclusions about the motions of galaxies.

In 1919, while Humason was ending his career as a mule skinner, the thirty-year-old Hubble had to choose an observing program—the key decision that any observational astronomer must make. The rising young man at the Mount Wilson Observatory, Harlow Shapley, had been born near the Ozarks of southwestern Missouri four years before Hubble came into the world seventy miles away. Shapley had gained prominence for his observations of variable stars in "globular clusters," compact objects that pack many thousands of stars into a region a few dozen light-years across. Surveying the sky with the 60-inch telescope on Mount Wilson, he had demonstrated that an inordinately large number of globular clusters lie in a particular direction, toward the constellation Sagittarius. Shapley concluded that the solar system belongs to a vast assemblage of stars, now called the "Milky Way galaxy," the center of which lies in the direction of Sagittarius. He rightly deduced that some of the globular clusters lie much farther from the center than the solar system does, so they appear in all directions on the sky, whereas a roughly equal number of globular clusters, much closer to the center than we are, display the concentration that he analyzed. Perhaps influenced by his concentration on the arrangement of objects in the Milky Way, Shapley proposed that *all* the objects detected by astronomers lie within it. The Milky Way certainly appeared to have a size much greater than that assignable to any other object or class of objects, implying a special role and position for our starry home.

17

(Right)
Harlow Shapley (1885–1972), born in Missouri like Hubble, played a key role in showing how Cepheid variable stars could be used to determine the distances to far away clusters of stars. (Photograph by Frank Hogg, courtesy of Helen Sawyer Hogg and Owen Gingerich.)

(Below)
Edwin Hubble (1889–1953) discovered that galaxies, each made of billions of stars, are moving away from our own Milky Way, with speeds that are proportional to their distances from us. This photograph shows Hubble pointing to the first Cepheid variable star he discovered in another galaxy, the famous Andromeda spiral. (Photograph courtesy of the Henry E. Huntington Library and Art Gallery.)

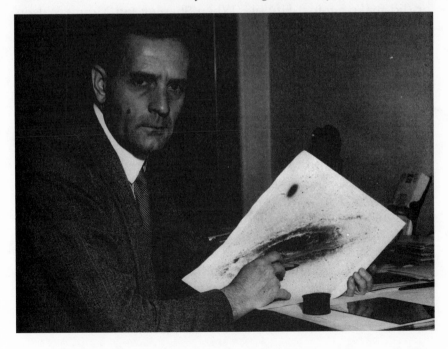

Despite, or because of, their Missouri background, Shapley and Hubble detested each other. Ancient astronomical lore, told to me by a much-beloved, long-deceased astronomer, assigns some of this hatred to an anonymous set of doggerel verses that appeared on the bulletin board at the observatory, referring to a woman whom Hubble was maintaining at the Mount Wilson Hotel, a popular local retreat not far from the telescopes. Hubble deduced the author to be Shapley, and the mountain might not have proven large enough to hold both astronomers peaceably, the more so as both astronomers wanted to study the most distant objects in the Milky Way and beyond (if anything lay beyond).

Fortunately for the greater glory of astronomy, Shapley had already impressed Harvard University as precisely the rising young star it needed to restore the Harvard Observatory to the highest rank, a position it had ceded, so far as professorial heavyweights were concerned, to its rival Princeton, where Henry Norris Russell, the dean of American astronomers (and Shapley's Ph.D. thesis advisor), presided. In 1921, eighteen months after Hubble joined the Mount Wilson staff, Shapley left California for Harvard, where he would succeed Russell as the most influential astronomer in the United States, eventually to be investigated by a congressional committee as a dangerous leftist, an accusation no one ever could or did make against Hubble. When Shapley left Mount Wilson, that investigation lay three decades in the future. The issue then on everyone's mind (not counting nonastronomers) provided the topic of the most famous debate in twentieth-century astronomical history, centering on the subject that Hubble had chosen for his own: the nature of the spiral nebulae.

Shapley's assertion that everything lies within the Milky Way had not gone unchallenged. Other astronomers insisted that the objects with intricate spiral patterns they had named "spiral nebulae" had approximately the same sizes as the Milky Way and lay far beyond its outermost confines, forming "island universes," as the phrase ran, in their own right. In order to decide whether spiral nebulae are subunits of our giant Milky Way or whether they stand on terms of equality with our own starry agglomeration, astronomers needed to know the size of the Milky Way and the distances to at least some of the spiral nebulae. They could measure the angular size of these spirals, but without any good idea of their distances from us, they could not determine whether they were

The galaxy M81, located in the constellation Ursa Major, was one of the "spiral nebulae," the nature of which dominated the Great Debate of 1920. We now know that M81 is a spiral galaxy similar to our own Milky Way. If this were an image of the Milky Way, the solar system would lie in one of the spiral arms, far from the galaxy's center in the photograph. (Photograph courtesy of the National Optical Astronomy Observatories.)

observing truly immense objects at enormous distances or merely huge objects at lesser distances.

THE GREAT DEBATE

Shapley argued for the latter interpretation. His chief opponent was Heber Curtis, a distinguished, somewhat older astronomer at the University of California's Lick Observatory, northern California's rival to the Mount Wilson Observatory. The controversy over the size of the Milky Way and the nature of the spiral nebulae, carried on in scientific publications, led the National Academy of Sciences to invite Shapley and Curtis to present their views in April 1920 at the academy's annual meeting in Washington, D.C. Astronomers always call the resulting session at the academy's meeting the "Great Debate," though they know that the evening in fact saw only the reading of consecutive papers by Curtis and Shapley. The debate derives its uppercase letters from the fact that Shapley and Curtis were struggling to determine the general arrangement of matter in the universe, an essential requirement for later attempts to discover the overall motions of that matter. In a career sense, less was at stake for Curtis, who already held a lifetime position at Lick Observatory, than for Shapley, who knew that the head of the Harvard committee to choose a new observatory director would be in careful attendance.[1]

Though Shapley had the wrong side in the Great Debate, he had the better supporting data. One of his and Hubble's colleagues at Mount Wilson, the Dutch-born astronomer Adrian van Maanen (who also despised Hubble for his Oxford affectations), had been photographing spiral nebulae for years. Careful measurement of his photographs convinced van Maanen, and many others, that the spiral nebulae were rotating, with specific parts of the complex spirals changing their positions slightly over a few years' time. The changes amounted to tiny shifts in angular measurements, not much greater than the minimum angle that the great telescopes could detect, but van Maanen had measured them.

These changes could be real only if the spiral nebulae lie relatively close to us. The velocities implied by the changes in the positions of pieces of the spiral nebulae must be larger if the distances were larger, since the nebulae would have had to have rotated over greater distances to produce these changes. Placing the spiral nebulae beyond the estimated extent of the Milky Way would imply that the spirals were rotating at speeds greater than the speed of light, conceded to be an impossibility by all involved. Curtis had to argue, more by implication than direct statement, that van Maanen's measurements were flat-out wrong, which indeed they turned out to be (though van Maanen never admitted this directly). Shapley, emphasizing van Maanen's work, had seized on incorrect data—a common danger in science, which scientists ignore at peril to their careers. As a famous scientific dictum states, "Beware the theory that agrees with *all* the data, for some of the data are wrong."

---

[1] A number of astronomical histories, including the standard biography of Hubble, Gale Christianson's *Edwin Hubble: Mariner of the Nebulae*, repeat Shapely's recollection that Einstein attended this debate, which would certainly have added further to Shapley's tension level. In fact, Einstein first visited the United States only in the following spring of 1921.

HUBBLE'S FIRST DISCOVERY:
CEPHEID VARIABLE STARS IN THE ANDROMEDA NEBULA

For several years, nothing could be proved definitively, but in 1923, Hubble ended the debate, making it clear in hindsight that Shapley was the loser. (Hubble, who must have been delighted at this outcome, was too much the gentleman to record his feelings on the matter.) Hubble had spent many a long night on Mount Wilson, photographing spiral nebulae with the 100-inch Hooker Telescope, completed in 1919 and named, not as many wags insisted, after either the Civil War general or the camp followers whom his easygoing ways had supposedly encouraged, but quite understandably after the philanthropist who had donated significant funds toward its construction. Without a totally fixed objective, Hubble strove to find some types of objects in the spirals that could be compared directly with a nearly identical object in our own galaxy, most preferably one whose distance had already been well determined. His persistence received its reward in 1923, when he found the first Cepheid variable star in the spiral nebula in Andromeda, and he soon followed this discovery with several others.

Cepheid variables had become well known to astronomers, most prominently through Shapley's efforts to use them in estimating the distances to globular clusters. We now know that Cepheid variables are high-mass stars nearing the ends of their nuclear-fusing lifetimes, with enormous intrinsic brightnesses, called "luminosities," that make them visible out to relatively great distances. What we might deem the uncertainty of old age causes the stars to pulsate; unable to find a true equilibrium, they alternately expand and contract, varying their energy output as they do so. Shapley's predecessors at Harvard—most notably, Henrietta Swan Leavitt—had demonstrated that Cepheid variables with greater average luminosities pulsate more slowly, while those with lower luminosities vary more rapidly. By observing a number of Cepheid variables, all at nearly the same distance from us in a nebula called the "Small Magellanic Cloud," Leavitt made the key discovery that every value for the period of light variation corresponds to just one average luminosity. Thus, if astronomers could identify two stars as Cepheid variables with the same pulsation period (as they could on the basis of the stars' variable light and relative intensities of the light with different colors), then they could justifiably conclude that the two stars have the same luminosity or intrinsic brightness. In that case, the Cepheid variable that seemed fainter must be farther from us. Because the brightness that we observe scales inversely with the square of the distance, we know that if one Cepheid variable has 100 times the apparent brightness of another with the same period of variation, then the fainter star must lie 10 times farther away.

The first Cepheid variable star that Hubble identified in the Andromeda nebula followed this rule, exhibiting an average apparent brightness only 1/100 of the brightness of the faintest Cepheid variables with the same period of variation that had been previously discovered. By a complex process of estimation, astronomers had already assigned distances of many tens of thousands of light-years to those Cepheids. Each light-year, the distance that light travels in a year, measures close to six trillion miles; the sun's closest neighbors have distances between four and five light-years. Hubble's Cepheid variable in the Andromeda nebula had to lie hundreds of thousands of light-years from the solar system, at least 100,000 times more distant from us than the closest stars.

Shapley had previously estimated the distance from the solar system to the center of the Milky Way at about 50,000 light-years and the full extent of the starry system at 300,000 light-years. No one believed that the Milky Way could really be larger than this, and quite a few astronomers felt that Shapley had puffed up his value to some extent. Now Hubble had found an object belonging to a spiral nebula at a distance at least twice as great as the size of the Milky Way. The news from California reverberated through the corridors of Harvard, Princeton, and Berkeley and crossed the ocean to Oxford, Paris, and Berlin: Spiral nebulae lie outside the Milky Way! Since the Andromeda nebula had the greatest apparent brightness and angular extent of all the spiral nebulae, by placing it beyond the Milky Way, Hubble effectively put them all outside. Before long, astronomers had dropped the term "spiral nebulae" and come to speak instead of "spiral galaxies." The word "galaxy," derived from the Greek word for milk, provides an ancient name for the Milky Way itself, and it now refers to all agglomerations of stars and gas roughly similar in size and mass to our own Milky Way, each of which contains many millions or billions of individual stars.

In addition to the spiral galaxies, some of the other types of neb-ulae also turned out to be galaxies. These include the elliptical galaxies, giant ellipsoids that each contain billions of stars, with al-most none of the interstellar gas and dust found in spiral galaxies. Ellipticals appear in space almost as frequently as spirals do. A small fraction of galaxies, designated as "S0," have the flattened disk shapes of spirals but lack spiral arms. Galaxies that are nei-ther spiral, elliptical, nor S0 belong to a mixed bag of smaller galaxies called "irregulars," less common than either of the two major types, of which the Large and Small Magellanic Clouds, which orbit our own galaxy as satellites, are the best-known exam-ples. The remaining objects that astronomers still label "nebulae" have deservedly retained their title, having shown themselves to later generations of astronomers as discrete, bloated masses within the Milky Way, either gaseous envelopes that aging stars have puffed or blown away, or much larger assemblages of gas and dust, ripe for contraction that will form a new generation of stars.

During the three decades of his life after 1923, Hubble stood without challenge as the acknowledged master of the distances to galaxies. In galaxy after galaxy, he discovered Cepheid variables and other objects useful in estimating distances. Hubble also cre-ated the classification scheme of galactic types, summarized in the previous paragraph, that remains the astronomical standard, even though modern discoveries have revealed more and more oddball additions to, and variations of, the three basic types. Just by estab-lishing the fundamental arrangement of matter into individual galaxies and recognizing the basic types of galaxies, Hubble would have become famous in astronomical history. But he achieved a still-greater impact with a simple graph, first published in 1929, that encapsulates what Hubble knew about galaxies' motions and had deduced about their distances. A discussion of this graph de-serves a chapter by itself, for it revolutionized human under-standing of the cosmos, rendering the cosmological constant unnecessary for the next sixty years.

CHAPTER FOUR

THE EXPANSION OF THE UNIVERSE

IN 1929, EDWIN HUBBLE PUBLISHED A DIAGRAM that summarized nearly a decade of observations, the fruit of his observational program undertaken to estimate the distances to two dozen galaxies. Working diligently through the previous half decade, Hubble managed to estimate galaxies' distances at the rate of about four per year. The slow pace at which his results emerged testifies to the difficulty of the work that Hubble pioneered: establishing the true scale of distances in the cosmos. As a result of his efforts, Hubble, who rather despised uncertainty and maintained a conservative style of life, complete with occasional speeches praising successful capitalists and movie stars, trusted his data. In the year that the stock market crashed, he published the diagram that changed human understanding of the universe.

THE QUEST FOR STANDARD CANDLES TO MEASURE DISTANCES IN THE UNIVERSE

Hubble's estimates of the distances to galaxies rested on the use of "standard candles," which astronomers define as a set of objects with identical luminosities (intrinsic brightnesses). Driving the desert highways on a clear night, you can see the lights of automobiles that are miles away, and you can roughly judge their distances by the apparent brightnesses of their headlights. This judgment relies on the assumption that all car headlights have the same luminosity, an assumption that could be wrong. When a car's headlights change from high to low beams, they change their luminosity by a factor of about four. Because the apparent brightness of any object decreases in proportion to the square of its distance from the observer, you will overestimate a car's distance by a factor of two if you think that its headlights are on high beams when in fact they are on low. Similarly, if you are observing not an automobile, but a truck with high beams that are twice as bright as an automobile's, you will underestimate your distance from the vehicle if you assume that you are observing light from a car, in this case by a factor equal to the square root of two.

Analogous problems confound astronomers' desires to find and to use standard candles for distance determinations. They must verify that all members of a class of objects have identical, or nearly identical, luminosities. Then they must find a way to identify new members of that class, typically seen at much greater distances, and therefore with much lower apparent brightnesses, than the previously known members. The next step seems simple: compare the apparent brightnesses of the bright and dim members of the class to determine how many times farther from us the dim members lie. In principle, this describes the method perfectly; in practice, however, a host of problems intervenes.

Astronomers require standard candles because they lack a direct means of measuring most of the distances that describe the cosmos. The single direct method for measuring the distances to stars employs the surveyor's approach of triangulation. As the Earth orbits the sun during the course of a year, astronomers look along slightly different lines of sight. In what scientists call the "parallax effect," the changes in the Earth's position make nearby stars appear to shift back and forth against the backdrop of more distant stars. By measuring the amounts of these shifts, astronomers can measure the distances to nearby stars through geometry: Stars closer to the solar system will display larger shifts. This method yields accurate distances to stars within a few hundred light-years of the sun—distances that are already tens of millions of times greater than the Earth-sun distance. For stars at still-greater distances, the parallax method simply fails, because instrumental errors and the blurring of starlight produced by Earth's atmosphere deny astronomers the chance of accurate measurements of the stars' tiny angular displacements.

FIGURE 4.1   The Parallax Effect

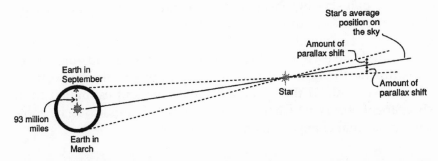

The parallax effect describes a star's apparent displacement, called its "parallax shift," against the backdrop of much more distant stars as the Earth orbits the sun during the course of a year. The size of the parallax shift varies in inverse proportion to the star's distance from the solar system. If this drawing were made to an accurate scale, and if the star under observation were one of the closest stars to the sun, more than a mile would separate the star from the solar system. (Diagram courtesy of Jon Lomberg.)

Hence astronomers need standard candles, such as the Cepheid variable stars that Shapley and Hubble used to such advantage. If they could use the parallax method to measure the distance to the closest of the Cepheid variables, all the others' distances could rest on this single measurement. The cosmos is not so kind: None of the Cepheid variables lies sufficiently close for the parallax effect to work.[1] Astronomers therefore have used a series of additional observational methods, over which we may draw the veil of simplicity, to estimate the distances to the closest Cepheids in the Milky Way. These stars serve as benchmarks, against which astronomers may compare the apparent brightnesses of much dimmer and more distant Cepheid variables.

By 1929, Hubble had employed Cepheids, along with other classes of objects thought to provide reasonably good standard candles, to derive distance estimates for about twenty galaxies beyond the Milky Way. He might have simply published a compilation of his work, gaining widespread admiration for his perseverance: Each individual Cepheid variable, for example, had

---

[1]This statement was absolutely true in Hubble's day. During the past few years, a European satellite named *Hipparcos* has vastly improved the accuracy of astronomers' observations of the parallax effect, extending the usefulness of this method by more than a factor of ten in distance and embracing, at the outer limits of its distance-determining ability, a few of the closest Cepheid variable stars.

to be discovered and followed photographically through many nights of painstaking observation, and the distances revealed by different types of supposed standard candles had to be cross-checked and reconciled. Hubble had intelligence, insight, and a way with the telescopes, which needed continual hands-on corrections during the hours of making long photographic exposures. When he examined the various distances he had estimated from six years of observing Cepheid variables in other galaxies, he spotted a trend that changed cosmology forever.

THE DOPPLER EFFECT REVEALS THE RECESSIONAL MOTIONS OF GALAXIES

How can a comparison of distances show that the universe is expanding? Hubble put together his distance estimates with results from the comparatively new science of spectroscopy, the investigation of objects by the details in their spectra, which amounts to studying objects by the colors of the light they produce. The spectrum of visible light extends from red light, which consists of light waves with the longest wavelengths and smallest frequencies of vibration, to violet light, with the shortest wavelengths and highest frequencies. This span describes what human eyes can detect; with laboratory equipment, we can observe waves with shorter wavelengths than those of violet light (ultraviolet, x rays, and gamma rays), as well as those with longer wavelengths than red light (infrared, microwave, and radio waves). Together, these types cover the full spectrum of electromagnetic radiation, which can reveal a great deal about the processes that produce radiation within a particular portion of the spectrum. For instance, the light from the sun and other stars typically shows a sudden dip in one particular shade of red light, the result of the absorption (blockage) of light by hydrogen atoms. Starlight also proves notably lacking in certain particular colors of yellow light, because calcium ions (atoms that have lost one or more electrons) in the stars' outer layers have absorbed this particular color of light as it began its journey outward.

While Hubble and Shapley strove to determine the spatial arrangement of the universe, other astronomers used spectroscopy to deduce that most of the features in the spectra of stars correspond to different chemical elements examined in terrestrial labo-

FIGURE 4.2   The Doppler Effect

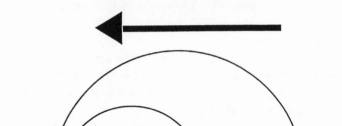

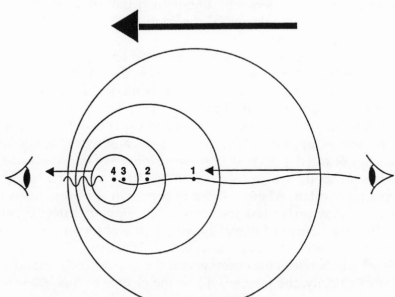

The Doppler effect arises when a source of wave motion (for example, of sound waves or light waves) moves with respect to an observer. The changing distance between the observer and the source either compresses the waves (if the source approaches the observer) or stretches them (if the source recedes from the observer). This compression or stretching appears as a change in the frequency (number of waves per second) and wavelength (distance between successive wave crests) of the waves that the observer measures. (Diagram courtesy of Jon Lomberg.)

ratories and that most stars have surface temperatures between 3,000 and 30,000 degrees Celsius. Late in the nineteenth century, spectroscopy had revealed one element never before seen on Earth, which astronomers named "helium," after the Greek word for the sun, *helios*; eventually, some helium would be found on Earth, trapped as gas in underground caverns. During the 1920s, as Hubble labored on his galaxies, the brilliant English-born astronomer Cecilia Payne, who had obtained a research position at Harvard but was long denied the professorship she richly deserved, analyzed the spectra of stars and demonstrated that most of them consist of nearly the same mixture of the elements, dominated by hydrogen and helium. Payne and those who followed her showed the astronomers who understood the colors of starlight how spectroscopy could reveal the composition of objects thousands of light-years away.

Spectroscopy offered one additional great boon to astronomers: No matter how distant an object might be, they could determine the object's speed toward or away from the observer who recorded the colors of its light in detail. The Doppler effect, named in honor of the scientist who first explained it concerning sound waves, appears whenever the source of waves either approaches or recedes from an observer. Motions of approach effectively squeeze the waves together, so that their crests and troughs arrive more often, whereas those of recession spread the waves apart, so they arrive less frequently. The Doppler effect occurs whether the source of waves moves, or the observer moves, or both are in motion; all that counts is the relative velocity along the line of sight between the observer and the source. The greater that relative velocity, the greater will be the change in the frequency (the rate at which successive waves arrive) and in the wavelength (the distance between successive wave crests) of the observed waves. All of the colors—that is, all of the frequencies and wavelengths—of light waves experience the same fractional changes as the result of the Doppler effect. Thus the Doppler effect for a velocity of recession preserves a pattern of alternatingly brighter and darker colors, while it reduces all the frequencies and increases all the wavelengths in that pattern. If astronomers observe a star with a familiar pattern in its spectrum, but if they find that all the wavelengths are 1 percent longer than usual and that all the frequencies are 1 percent lower, they conclude that the star is moving away from us at 1 percent of the speed of light. They cannot say, however, which object (or both) partakes of this motion; only the relative velocity matters for the Doppler effect. Nor can they determine how the object is moving across our line of sight: The Doppler effect reveals information only about relative motion toward or away from the observer.

Hubble was no expert in spectra, but he knew the results obtained by those who were, most prominently Vesto Slipher, an astronomer working at the Lowell Observatory near Flagstaff, Arizona. This observatory, founded by Percival Lowell to study one of the closest celestial objects (the planet Mars), provided basic data for understanding the cosmos as a whole. During the first decades of the twentieth century, Slipher had obtained many spectra of the brightest galaxies (as they turned out to be), including the most prominent spirals. He saw that the colors of light in these galaxies followed patterns already familiar from studies of stellar spectra, proof that galaxies owe their light to the combined contributions from billions of stars. Slipher also saw that the patterns of light had been shifted, in some cases to shorter wavelengths and higher frequencies but usually to longer wavelengths and lower frequencies. Astronomers had begun to speak of the former as "blue shifts," meaning shifts to shorter wavelengths, and of the latter as "red shifts," indicating shifts to longer wavelengths. Careful measurement disclosed the amounts of these shifts, and thus the speed at which blue-shifted galaxies were approaching ours and red-shifted ones receding from us.

HUBBLE DISCOVERS THE EXPANDING UNIVERSE

By taking these data on galaxies' motions and combining them with his own on galaxies' distances, Edwin Hubble changed our understanding of the cosmos. He did what scientists do naturally, taking two or more sets of data that describe a group of objects in different ways and searching for correlations that may reveal an underlying truth. The most familiar means of displaying data to uncover any correlations uses a graph to plot the objects' characteristics. Once Hubble had published his graph of galaxies' distances and velocities, astronomers soon came to realize that a new era in cosmology had been born.

Hubble's first graph, published in the *Proceedings of the National Academy of Sciences* in January 1929, showed the velocities of galaxies in the vertical direction and their estimated distances along the horizontal axis (see page 33). Today most astronomers draw such a diagram with the axes interchanged—that is, with distances plotted vertically and velocities horizontally. The choice reflects only matters of taste, though astronomers may have been influenced by the fact that the velocities have much smaller observational errors associated with them than the distances do. Galaxies' recession velocities can often be measured to plus or minus 1 percent, or even better, whereas even today the distance estimates for galaxies embrace errors of at least plus or minus 5 percent. In fact, Hubble's distance estimates were far too small, not by 5 or 25 or 50 percent, but by multiples of four to eight!

How, then, did Hubble ever manage to discover the expanding universe? His diagram shows that a few of the closest galaxies actually have velocities of approach. (These galaxies may not be headed straight for us; since astronomers can measure only velocities along the lines of sight to the galaxies, and not in the perpendicular directions, we can say only that the galaxies' motions are bringing them closer to us rather than carrying them farther away.) Most of the galaxies that Hubble plotted on his graph are receding from us, a rather mundane fact in view of the restricted number of galaxies whose distances he could estimate. What counts in the diagram is the straightness of either of his two lines, each of which shows the relationship that Hubble discerned between the galaxies' distances and their recession velocities. The two straight lines in the diagram reflect the fact that Hubble was not sure whether or not to include all the data he had assembled. The dashed line shows the relationship that he derived after omitting some of the more questionable observational results.

The relationship between velocities and distances gives Hubble's diagram its cosmic significance. Even though Hubble underestimated all the galaxy distances, he underestimated them in nearly the same proportion. As a result, the fundamental relationship Hubble discerned—that galaxies' recession velocities increase in proportion to their distances from us—remains valid, even after all the distances have been heavily revised. Today astronomers refer to the relationship between distances and velocities as "Hubble's law," expressed algebraically as $v = H \times d$. Here v represents velocity of a galaxy along our line of sight, d stands for its distance, and H is a constant, known as the "Hubble constant," that numerically captures the proportionality between v and d.

THE COSMOLOGICAL PRINCIPLE

If we accept the validity of the velocity-distance relationship that Hubble discovered, why does this prove that the universe is expanding? Why should we not conclude, for example, that galaxies are indeed moving away from our neighborhood but are simultaneously coming together somewhere else, converging toward a point that lies beyond our observational abilities? All of modern

FIGURE 4.3   Hubble's First Velocity-Distance Diagram

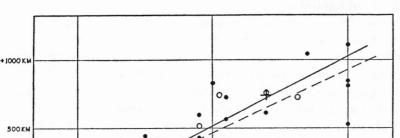

*In 1929, Edwin Hubble published the original version of what as-
tronomers now call the "velocity-distance diagram" or the "Hubble diagram." This graph shows the distances to nearby galaxies along with their
velocities, almost all of which are velocities of recession (greater than zero)
rather than of approach. Hubble misleadingly labeled the velocity units
along the vertical axis as "KM" (kilometers) rather than kilometers per
second. These observations extend only to distances estimated at 2 million
parsecs, approximately 6.5 million light-years, less than 1/1,000 of the
distances out to which astronomers can now observe galaxies.* (Diagram
from the *Proceedings of the National Academy of Sciences,* courtesy of
the National Academy of Sciences.)

cosmology rests on the assumption that we have a representative
view of the universe. This hypothesis has no basis in observational
reality, other than the fact that the cosmos that we see does look
much the same in all directions. We might, however, be fooled by a
limited horizon of observation, like local philosophers drawing
general conclusions about humanity from the few people they
know. The principle that what we see provides us with a representative slice of reality can never be fully verified. Astronomers support this hypothesis, which they call the "cosmological principle,"
because it offers the most straightforward statement about the cosmos that we cannot see. Yet they remain fully aware that the cos-

mological principle rests on what cosmologists choose to hypothe-size about the universe.

If we join in this hypothesis, we can admire the predictive power of the cosmological principle, which implies that every observer, everywhere in the universe, should observe what we do: Galaxies are receding from that observer according to Hubble's law, with the same value of the Hubble constant that we derive from our ob-servations. If all observers see galaxies in recession, with the galax-ies at greater distances receding more rapidly, then the entire universe must be expanding, as the basic agglomerations of matter move apart throughout the cosmos.[2]

MODELS OF THE EXPANDING UNIVERSE

How can this be so? Can the cosmos really expand everywhere, with all observers recording the same basic pattern? Doesn't the cosmic expansion imply a central stationary point, with all galaxies receding from it, that must be different from the other locations in the universe? Hubble's observations plus the cosmological princi-ple provoke these questions, which turn out to have complete, if not completely satisfying, answers. Astronomers who attempt to ease the public into accepting the idea of an expanding universe start by urging acceptance of the fact that no one can truly imagine the cosmos. To hold the universe in mind while contemplating its expansion demands more than human brainpower—not in capac-ity, but in conceptual ability. By definition, the universe contains all that exists, and we cannot step outside it, even in our minds, to ob-tain a clear view. Instead, we must make do with mathematics, models, and approximations. These serve us well, but they will never fully spread the embroidered cloths of the heavens before us.

The favorite model of cosmologists reduces three-dimensional space to the two-dimensional surface of an expanding balloon, so that we can indeed step outside the model and watch it expand. If we glue stickers on the balloon to represent galaxies and insist that light can travel only around the balloon's surface, we can verify that as the balloon expands, every sticker will see all the others re-

---

[2]A distinction can be made between the basic Doppler effect, which arises from objects' motions with respect to one another, and the shifts in the spectra of the light from distant galaxies, which arise from the expansion of space itself. Because this distinction lies well be-yond the scope of this book, it is simply noted here; those who wish to pursue it further may consult Edward Harrison's *Cosmology: The Science of the Universe*.

ceding. Indeed, the stickers obey Hubble's law: A sticker twice as distant as another from any reference point will move away twice as rapidly. No sticker can claim to be the stationary center of the expansion. If a center exists, it lies not on the balloon's surface, but in its interior, which does not exist in the universe that the model embodies. (Indeed, cosmologists can describe the expansion of three-dimensional space as proceeding around a point located in a fourth dimension that exists only mathematically.)

The balloon model of the expanding universe also helps to resolve another objection to the concept of an expanding universe: What is space expanding into? In the model, only the balloon's surface exists, representing all of three-dimensional space, and what we take to be space represents merely a mathematical construct. Thus the balloon can expand, gaining a larger surface area that stands for a greater total amount of space, without running into anything. Those who object to this sleight of space must remind themselves that no model can furnish a wholly satisfactory explanation of how the cosmos behaves.

Finally, the balloon model provides a particular representation of the cosmos, one in which the total volume of space always remains finite. The balloon's surface will always include only a finite amount of surface, no matter how long the expansion continues. After an infinite amount of time, to be sure, the surface area will become infinite, but that requires quite a wait; during any finite time interval, no matter how large, the surface area, which stands for the total amount of space in the universe, will stay finite. Theoretically at least, we could measure this total volume, and an astronaut with endurance could set out on a journey in a straight line that would eventually bring her back to her starting point.

We can also imagine a universe with an infinite amount of space, and we can model it as the surface of an expanding rubber sheet. For most people, this model seems more reasonable, but notice that it basically takes the conceptual problem of what space expands into and pushes it out to infinity, where it seems to disappear. The crucial point about the flat model of an infinite cosmos is not that it may seem easier to understand, but rather that it represents a realistic possibility, just as the finite model does.

A third category of models represents space with what cosmologists call "negative curvature," in contrast to the "positive curvature" of space in a finite, balloonlike universe. To model negative curvature in two dimensions, we must imagine a surface that at every point resembles the surface of a saddle, so that space curves in one sense if we move toward the horse's head or tail and in the opposite sense if we move toward the stirrups. In its most significant feature, negatively curved space resembles flat space: Both extend to infinity and include an infinite volume.[3]

The models described above have a long and honorable history, providing two-dimensional analogues to real, three-dimensional space. Well before Hubble published his epochal paper in 1929, cosmologists had used Einstein's equations to determine the possible forms that the universe might take. First among these efforts was Einstein's own, in which a cosmological constant allowed a static universe to exist. Next, a month after Einstein published his cosmological-constant theory in February 1917, the Dutch cosmologist Willem de Sitter showed that Einstein's equations allowed the existence of a static model that differed from Einstein's: This universe was completely empty, utterly devoid of matter! If, however, even a few particles of matter were introduced into such a universe, it would immediately start to expand and would continue expanding at an ever-increasing rate, driven by a cosmological constant with a value different from the one Einstein had envisioned. We shall meet the de Sitter model later in this book in two separate contexts. Although the model clearly does not describe the universe today, it provides a good description of the cosmos at two utterly different stages of development: during the inflationary era, a moment after the universe began, and in the far-distant future, if the cosmological constant does not equal zero.

The last paragraph of Hubble's original paper refers to "the possibility that the observed velocity-distance relation may represent the de Sitter effect." This underscores the looseness of Hubble's links with theoretical cosmology: He would have more appropriately referred to models of the cosmos that had been developed in 1922 by Alexander Friedmann in the Soviet Union and in 1927 by Georges Lemaître, a Catholic abbé in Belgium. Friedmann's work, like the man himself, who died young, a victim of the poor living conditions in the Soviet Union, passed almost unnoticed, but Lemaître's received wide circulation among cosmologists shortly before Hubble published his work on the velocity-distance relationship.

---

[3]Mathematicians have devised equations describing negatively curved space that, like positively curved space, includes only a finite volume. We should therefore include finite, negatively curved space in our list of possible universes.

Both Friedmann and Lemaître achieved what Einstein just missed: the discovery, based on nothing but equations, that the universe may be expanding. The cosmological models they independently created, now called "Friedmann-Lemaître models," dealt with a universe in which the cosmological constant equals zero (Friedmann made this an absolute condition, whereas Lemaître was willing to allow a nonzero value). Unlike Einstein, Friedmann and Lemaître admitted the possibility that space exists in a state of expansion or contraction. This led them to create a set of models that differ in the curvature of space, which, as we have seen, can be either positively curved, negatively curved, or flat. If the cosmological constant equals zero, the curvature of space depends directly on the average density of matter. Any model of an expanding universe must have begun its expansion at a particular moment in time, now called the "big bang."

Hubble's discovery of the velocity-distance relationship, combined with the cosmological principle, implies that one of the Friedmann-Lemaître models represents the real universe, with space expanding away from every point in the cosmos. We must reject the temptation to imagine that space simply sits while objects move through it, and we must instead attempt to bend our minds to the notion that space itself expands. Otherwise, the "expansion of the universe" would simply amount to a great rush of matter through constant space. Einstein's equations, and all that we know about the cosmos, imply that we must reject this intuitively obvious concept in favor of space that itself expands. We might pause to ask ourselves, Is space something, or is it nothing? If it is nothing but the distance between objects, then space, by definition, expands as objects move apart. If space is something, what is it? A full answer lies, to put it mildly, beyond the confines of this book, which can achieve its objective by urging readers to examine their beliefs about the nature of space and by stating from authority that if we hope for an accurate understanding of the universe, we must imagine that space—whatever that may be—expands as the universe expands.

Though the larger world had other things on its mind, the scientific community took immediate notice of Hubble's work in 1929. In Berlin, Albert Einstein must have been deeply impressed. He had, of course, followed Hubble's discovery that the spiral nebulae are other galaxies and must have enjoyed learning that the universe is expanding. When Einstein visited California in the winter of 1930–1931, he spent considerable time at the California Institute of Technology and made the trip up Mount Wilson to visit Hubble. There he pronounced his acceptance of the expanding universe and, by implication, abandoned as useless the cosmological constant he had invented fifteen years before. The expanding universe had turned Einstein's constant into a blind alley; what counted for the future was extending Hubble's original diagram deeper into space.

From Hubble's First Results Toward an Ever-Larger Universe

Two years after Hubble's original paper appeared, he and Milton Humason published new observations of galaxies, extending the straight-line relationship between distances and velocities by a factor of twenty. With this 1931 paper, Hubble and Humason overcame the understandable skepticism concerning Hubble's earlier, preliminary results. Thereafter, the expanding universe became a fixed point in the constellation of astronomical thinking. Two issues stemming from Hubble's law and the cosmological principle then seemed paramount: How long has the expansion been going on? And will the expansion continue forever? Hubble's law partially answers the first of these questions, Einstein's equations the second.

Imagine the history of the universe to be captured on film, which we can run backward if we choose. Since all galaxies are moving away from one another, they must have been closer together in the past. Further back in time, they must have been still closer to one another. If we roll the movie still further back, we come to a time when all the galaxies were next to one another. More precisely, since galaxies themselves have evolved with time, we come to epochs when all the matter in the universe crowded together at densities that increased without limit. (Notice, for mind-bending purposes, that if the universe contains an infinite amount of space, then even at these moments, space was infinite—yet the density of matter nevertheless rises without limit as we look backward in time.) The movie begins at a precise instant in time, the moment of the big bang. Thus the question of how long the expansion has been going on becomes, to astronomers, the issue of locating the big bang at a particular time in the past.

Hubble's law can do this, at least in approximation. To see why this is so, we do best to write Hubble's law in reverse order, as H x d = v, and then to divide both sides of this relationship by H. The equation then becomes d = v x (1/H). To physicists, this equation oozes with significance. If we assume that throughout the history of the universe, no deceleration or acceleration has occurred, then the velocity v of any galaxy must have remained constant. In that case, Hubble's law tells us that the distance at which any representative being observes a particular object must equal the constant speed, v, at which the object has been receding, multiplied by 1/H. But we know that if objects receding at a constant velocity from an observer began their motion at zero distance, the distance that they have now reached equals that velocity times the amount of time that they have been in motion. Thus 1/H must equal the time since the expansion began, provided that galaxies' recession velocities have remained unchanged throughout history.

THE QUEST TO DETERMINE
THE HUBBLE CONSTANT

In fact, despite Hubble's years of hard work, all his distance estimates had grievous errors, because astronomers had mistakenly assumed that the closer Cepheid variable stars in the Milky Way were identical to those with the same period of variation observed in other galaxies. Twenty years later, astronomers recognized that Cepheids come in two distinct types, which have different intrinsic brightnesses for the same period of cyclical light variation. Once astronomers learned how to use the colors of these stars to distinguish one type of Cepheid from the other, they reexamined Hubble's distance estimates and found them all to be too low by factors of four to eight.

Several important lessons lurk in this story. First, as mentioned earlier, because Hubble underestimated all the distances to other galaxies by approximately the same factor, the comparison of distances—the fact that certain distances are two or three or five times others—has proven nearly correct. It is not the absolute values of the distances, but this *comparison* of distances that reveals the expansion of the universe. We need to know the actual value of 1/H in order to determine how long the universe has been expanding, but the fact that v = H x d implies a universal expansion, even if we derive a mistaken value for H.

Hubble's initial work gave a value for H that we now know to be much too large. Hence the value of 1/H that Hubble found was much too small—less, indeed, than the age of the Earth, which was already known to have a value greater than two billion years. Later work sharply increased the estimated distances to galaxies. This decreased the value of Hubble's constant H: If v = H x d, to obtain the values of v observed for individual galaxies from a set of *larger* distances would require a *smaller* value of H. As the value of H fell, that of 1/H rose, and so did the age of the universe. By the mid-1950s, astronomers' best estimate of 1/H had reached a comfortable 19 billion years, several times the age of the solar system and older than the estimated age of any stars. During the early 1990s, a flurry of panic arose when still more refined observations raised the value of H somewhat, threatening to reduce 1/H to 10 or 12 billion years, an amount close to, or even less than, the estimated ages of the oldest stars. Although one escape route from this problem lay in adopting a nonzero value of the cosmological constant, cooler heads waited for yet more improved observations, which lowered H and raised 1/H. The best estimates now set 1/H at 15 billion years, give or take a billion. Since the oldest stars have ages estimated at 11 or 12 billion years, everything seems to fit, even though we must allow for some slowing down of the expansion, which may make the age of the universe 1 or 2 billion years less than 1/H. The rocking of the cosmological boat aroused by the recent supernova observations arises from the apparent discovery of a nonzero cosmological constant, while the deduced age of the universe has settled rather comfortably at a value that almost all astronomers find acceptable.

Modifying the Hubble Diagram for Modern Assaults on Cosmology

Hubble wrote his law, quite understandably, as v = H x d, velocity equals a constant multiplied by distance. He knew, of course, that Slipher and other astronomers measured not the velocities of galaxies, but rather their redshifts, the fractional amounts by which all the wavelengths in the spectrum of each galaxy's light had increased. These redshifts must be converted into velocities in accordance with

our understanding of the relationship between the galaxy's reces-
sional velocity and the redshift observed in its spectrum. For velocities much less than the speed of light, this relationship takes the
utterly straightforward form $z = v/c$, where z is the fractional increase in all the wavelengths, v is the velocity of recession, and c is
the speed of light. Converting from redshift to velocity then involves the simple act of multiplying z by c, the speed of light.[4]

For velocities that rise to a fair fraction of the speed of light, Einstein's special theory of relativity produces a more complex relationship (provided in Note 4 below). As astronomers began to
observe galaxies so far away that their recession velocities reach
astounding numbers, they found it more convenient to refer simply to each galaxy's redshift, which, after all, is what they observe,
and to leave the conversion into a recession velocity as an exercise
for their students. All modern versions of the Hubble diagram plot
not velocity versus distance, but rather redshift versus distance, or,
more precisely, redshift versus a measure of the distance.

We know from Einstein's equations describing the expansion of
the universe that whenever astronomers measure the redshift in a
galaxy's spectrum, they obtain a useful measure of the time after
the big bang when the galaxy's light left on its journey toward us.
However, this time appears not as an absolute number, measured
in years, but *as a fraction of the time since the big bang*. For a flat universe, this fraction equals $1/(1 + z)^{3/2}$ (i.e., one over the three-halves
power of $[1 + z]$), where z is the redshift—the proportion by which
the wavelengths in the spectrum have increased over the values
they would have if the redshift were 0. If z equals 0, the fraction of
time since the big bang equals 1, which seems reasonable. For $z =$
1, this fraction equals $1/2.8$, so the light from a galaxy whose
wavelengths have all doubled shows the galaxy as it was when the
cosmos had $1/2.8$, or just over 35 percent, of its present age. If $z =$
3, all the wavelengths have become four times longer than they
would be if no recession existed, and we see the galaxy as it was
when the universe had $1/8$ of its present age. As astronomers compete for the temporary distinction of finding the object with the

---

[4]Einstein's special theory of relativity shows that the redshift z plus 1 is equal to the
square root of the ratio $[1 + (v/c)]/[1 - (v/c)]$. Note that if v equals 0, this ratio equals 1, so z
equals 0. As (v/c) climbs close to 1, z can increase to values many times larger than 1 because the denominator in the ratio, $[1 - (v/c)]$, becomes extremely small. With algebra, we
can show that (v/c) equals the ratio $[z^2 + 2z]/[z^2 + 2z + 2]$. This ratio equals 0 when z equals
0 and never rises all the way to 1, no matter how large z becomes.

greatest known redshift, they may yet reach a redshift z = 8, which refers to galaxies whose light left when the universe had 1/27 of its present age. We are not there yet: The current high-redshift record belongs to a galaxy with a redshift of 5.6, which we see as it was when the universe was only 6 percent as old as it is now. These fractions are completely accurate only if space in the universe is flat. In other cases, a more complex relationship exists between an object's redshift and the time when the light we now observe left on its journey, though the ratio $1/(1 + z)^{3/2}$ provides a rather good approximation.

Astronomers can state the percentages of the universe's present age with a high degree of accuracy because they can measure redshifts with precision. Provided that we know the curvature of space, each redshift corresponds to a particular fraction of the universe's present age. To specify an exact age for the entire universe—the total elapsed time since the big bang—has proven more difficult than to measure galaxies' redshifts. Astronomers therefore prefer to talk in terms of the redshifts and to discuss the galaxies' corresponding ages (as we see them) as fractions of the total time since the big bang, rather than in numbers of years. For the public, astronomers multiply these fractions by their best estimates of the age of the universe, about 14 billion years. This allows them to state, for example, that the light from a galaxy whose redshift equals 5.6 left when the cosmos was less than 900 million years old. For objects observed at the greatest distances, the distance to the object, measured in light-years, rises almost to the age of the universe, stated in years. Designating the actual distances then involves the same uncertainty as that arising from attempts to measure the age of the universe. In contrast, astronomers can specify, with the same degree of accuracy with which they can measure a galaxy's redshift, the small fraction of the universe's present age that the cosmos had when the galaxy produced the glow that we now observe.

DISCOVERING THE FUTURE OF THE UNIVERSE BY LOOKING BACK THROUGH TIME

If we can determine H accurately, Hubble's law will tell us the age of the cosmos, provided that the rate of expansion has not changed. But Einstein's equations show us that sufficiently accurate observations of faraway galaxies will reveal the *future* of the universe. If astronomers can observe how the expansion has changed through past eras, they can use this knowledge to predict how it will change in the future—in particular, whether or not the expansion will ever cease, possibly to reverse into a universal contraction. If such a contraction occurs, it will eventually produce a "big crunch," in which the density of matter rises to an enormous value, reminiscent of conditions soon after the big bang. Indeed, we may speculate (with no deep justification) that a big crunch might recycle all space and matter through another big bang.

But how can the history of the universe reveal our cosmic future? The Hubble diagram looks backward in time by looking outward in space. More precisely, each point on the Hubble diagram represents the redshift and distance of a particular object. Because light takes larger amounts of time to cover increasing amounts of space, the light from objects at larger distances have longer "look-back times": We see them as they were further back in the past. Because of the different look-back times for galaxies at various distances, the Hubble diagram provides us with a view of the universe's history—not a snapshot of any one epoch, but a palimpsest, an overlay that displays objects whose light has taken different amounts of time to reach us and which we therefore see as they were at different eras in the history of the universe.

The extreme lower left-hand portion of the Hubble diagram (Figure 4.4)—the part that Hubble and Humason first investigated—represents objects with small redshifts and distances. These galaxies and galaxy clusters lie so close to us (no more than a few hundred million light-years away) that this portion of the graph essentially refers to the present era (more precisely, to times no more than a few hundred million years ago). These galaxies obey the simplest version of Hubble's law, with recession velocities proportional to their distances. Furthermore, because all their recession velocities are much less than the speed of light, the galaxies' redshifts are directly proportional to these velocities. From this portion of the diagram, astronomers may hope to obtain the present value of H. They must, however, distinguish the value that the Hubble constant has today from the values that it had in the past and will have in the future. As the expansion slows down or speeds up, the Hubble constant must change to reflect a new relationship between galaxies' distances and recession velocities. Astronomers therefore write H_0 to designate the value of the Hubble constant today. To be sure, by "today" they mean something like "within half a billion years of the present time."

FIGURE 4.4   The Modern Velocity-Distance Diagram

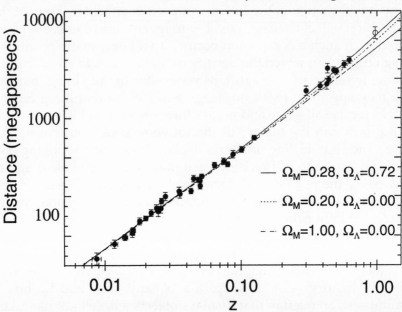

A recent version of the cosmological velocity-distance diagram plots the distances of galaxies along the vertical axis and the galaxies' redshifts along the horizontal axis. For galaxies receding at speeds far less than the speed of light, the recession velocities are directly proportional to the redshifts that astronomers measure in the galaxies' spectra. The straight-line relationship between distance and velocity (or redshift) that Hubble discovered holds true for galaxies thousands of times more distant than those that Hubble first examined. The crucial upper-right-hand portion of the diagram is analyzed in Chapter 7. (Diagram courtesy of Dr. Adam Riess and the High-Z Supernova Search Team.)

If the cosmological constant equals zero, astronomers have long known how the Hubble diagram should look as they observe objects at ever-greater distances. Different models of the cosmos, each of which depends on the average density of matter in the universe, trace out different lines on the Hubble diagram as we look toward the upper right, where recession velocities begin to rise close to the speed of light, and distances are measured in many billion light-years. No more than one of these lines can represent the actual universe. For generations, cosmologists have seen their task as finding the line of truth in the Hubble diagram that delineates our universe, not an imaginary model. Figure 4.4 shows two model universes (denoted by $\Omega_\Lambda = 0$) with a zero cosmological constant, one of which has an average density of matter equal to the critical density ($\Omega_M = 1.0$) and one of which has $\Omega_M = 0.20$, meaning that the average density of matter equals one-fifth of the critical value that would cause the universe to contract.

Unfortunately, the observational data do not yet isolate one single model as the real universe. Astronomers typically show the estimated inaccuracies in the distance determinations as "error bars" extending above and below the likeliest value for the distance to each object. Although some errors also arise in measuring the objects' redshifts, these errors are so small (typically no more than 1 or 2 percent) in comparison with the errors in the distance estimates that astronomers can safely ignore them without lessening the accuracy of their conclusions. Chapter 7 describes the startling results from astronomers' latest attempts to describe the universe with an accurate version of the Hubble diagram. In order fully to appreciate these efforts, we must define and explore the two crucial parameters that determine the actual line that the universe will take in this graph of redshift versus estimated distance.

THE TWO KEY COSMOLOGICAL PARAMETERS THAT DETERMINE THE FUTURE OF THE UNIVERSE

In addition to H_O, which specifies the universe's present rate of expansion, cosmologists seek two crucial numbers, each of which is designated by Ω, the Greek capital letter omega. One of these, Ω_M, describes the average density of matter in the universe. The other, Ω_Λ, characterizes the effect of the cosmological constant on the expanding universe; the cosmological constant is usually designated by Λ, the capital Greek letter lambda. Let us first consider the importance of Ω_M under the happy supposition, which astronomers held for decades, that Ω_Λ equals zero because the cosmological constant equals zero. Then, with a firm grasp on the implications of Ω_M, we can return to Ω_Λ in order to solidify our understanding of the three great numbers of the cosmos.

The number Ω_M describes the average density of all the matter in the universe in terms of what cosmologists call the "critical value" of the density; that is, Ω_M equals the *ratio* of the actual value of the density to the calculated critical value. This critical value equals a constant number times the square of the Hubble constant.[5]

The critical value of the density owes its name to the crucial role that it plays in determining the geometry and fate of the universe. If the cosmological constant equals zero, then the universe is positively curved and will eventually contract, *if and only if the actual value of the density exceeds the critical density.* Conversely, if the actual density falls below the critical density, the universe must be negatively curved and will expand forever. If the actual density exactly equals the critical density, space in the universe must be flat, and the universe will expand more and more slowly as time goes on, but it will cease its expansion completely only after an infinite amount of time has passed.

In all three cases, the expansion of the universe reduces the average density of matter, because the same amount of matter occupies a greater volume of space. This might lead one to think that although the average density may exceed the critical value for the density at some time, it must eventually fall below the critical density at some point as the universe continues to expand. However, the critical value for the density also changes with time, because it is proportional to the square of the Hubble constant H. If H changes, so will the critical density. The upshot is that if the actual density exceeds the critical value at any time, it will do so for all time. Likewise, if the actual density ever has a value less than the critical value for the density, it will always do so.[6]

Once cosmologists had perceived these facts, soon after Einstein and others analyzed the basic equations that describe the universe's expansion, an ever-growing desire burgeoned within their breasts to determine Ω_M, the ratio of the actual density to the critical value. In a universe with a zero cosmological constant, the value of Ω_M specifies the future of the universe, which must eventually contract if Ω_M exceeds 1 and will never stop expanding if Ω_M is less than or equal to 1. In order to determine Ω_M we must recognize that the existence of matter in the universe tends to slow the

---

[5]For those fond of algebra, the critical value of the density at the present epoch equals $3H_O^2/8\pi G$, where G is the universal constant of gravitation, found in Newton's law of gravity, and π is the familiar Greek letter pi, which represents the ratio of a circle's circumference to its diameter, approximately 3.14. At any time, the critical density equals $3H^2/8\pi G$, but H may have a value that differs from its value today.

[6]This statement holds true in a universe in which the cosmological constant equals zero, but not if the constant has a nonzero value.

expansion, as all objects with mass attract all other such objects through gravitational forces. When we look back in time, we expect to find that the universe had a greater rate of expansion than it does now. The question becomes, How much greater? In other words, how much has the expansion slowed down during the past? Greater amounts of slowing down—"deceleration" in scientific parlance—imply greater amounts of mass, that is, a larger value for the average density of matter.

To observe the slowing of the universal expansion, we must look far out in space, hence far back in time. As we have seen, astronomers execute this project by constructing a Hubble diagram that extends to these enormous distances. For four decades, from the early 1950s to the early 1990s, astronomers strove to derive distance estimates to faraway galaxies with an accuracy sufficient to show just which line through the upper right-hand portion of the Hubble diagram describes the real universe. These attempts all failed: Estimating the distances to galaxies many billion light-years away proved so difficult, with each estimate weighed down by large and unavoidable errors, that astronomers could draw no firm conclusions about the change in the expansion rate, and thus about the value of Ω_M, except to say that Ω_M could not be much larger than 1 or much smaller than 0.2. But on the crucial question of whether Ω_M exceeds 1 or falls below it, the Hubble diagram maintained an inscrutable silence, awaiting more accurate observations that would reveal which line on the diagram marks the real universe.

All these attempts to determine the future of the universe rested on the assumption that Ω_Λ equals 0 because Λ, the cosmological constant, equals 0. Eager though the reader may be to explore the new realms of possibility opened by a nonzero cosmological constant, if we hope to profit maximally from this tour, we must first make an excursion through two great breakthroughs of pre-cosmological-constant cosmology: the inflationary model and the discovery of dark matter in the universe.

CHAPTER FIVE

THE INFLATIONARY COSMOS

TWENTY YEARS AGO, DURING WHAT WE might now consider the late medieval period of modern cosmology, astronomers who hoped to resolve the mysteries of the universe had already acquired four fundamental pieces of cosmological information. First, the observed recession velocities and estimated distances of galaxies, when coupled with the assumption that we have a representative view of the universe, had demonstrated that the universe is expanding. The precise relationship between galaxies' distances and their recession velocities had revealed the value of the Hubble constant, and thus the approximate time since the big bang, the moment when the expansion began. This extent of time could (narrowly!) accommodate the greatest ages estimated for stars and galaxies, without requiring a cosmological constant. Second, the cosmic background radiation, discovered in 1964 and described in Chapter 11 of this book, had confirmed the validity of the big-bang model and had demonstrated that an "era of decoupling" of the radiation from the matter had occurred about 300,000 years after the big bang. Third, the hosts of visible galaxies offered ocular proof that what had been a smooth and nearly homogeneous universe soon after the big bang had somehow evolved into a cosmos within which matter had become highly clumped. Finally, the measured abundances of different varieties of hydrogen and helium matched astrophysical theories of how nuclei had fused together during the first minutes after the big bang.

During the 1970s, these observational data fit together rather well, so that all that seemed lacking in cosmology was an accurate determination of the average density of matter. This number, expressed by the parameter Ω_M, would determine whether the universe would expand forever. As described in the next chapter, astronomers had redoubled their efforts to detect and to measure different sorts of matter in different forms, never losing sight of the importance of comparing Ω_M with 1 to find the fate and shape of the universe. Whatever the value of Ω_M might be, the basic features of the cosmos and its history, which cosmologists called the "standard big-bang model" of the universe, apparently rested on firm observational and theoretical ground. Einstein's cosmological constant had received a decent burial by Einstein himself, and the occasional attempts to galvanize its corpse had seemingly proven nugatory. Yet a few clouds no bigger than a man's hand threatened the mental peace and quiet of thoughtful cosmologists.

Even though the observational data did not allow the determination of the crucial parameter Ω_M, they confronted cosmologists with a set of paradoxes, of which the two most important came to be called the "flatness problem" and the "horizon problem" of the universe. The great cosmological advances of the early 1980s came not from new observations, but rather from a new theory, one that dealt with the earliest moments after the big bang in a way that could resolve these paradoxes. This theory, motivated by new discoveries in the realm of particle physics, acquired the name of the "inflationary universe," or "inflation" for short. The inflationary model of the cosmos involves such mind-bending assumptions that no one accepted it on first reading, but it has exerted an immense appeal on cosmologists, at least on those of a theoretical bent. Before we examine this model, assess its claims to validity, or ask what observational data could verify or disprove it, we must motivate ourselves, as cosmologists did, by considering the paradoxes that the inflationary model can circumvent.

THE FLATNESS PROBLEM

The first of these paradoxes consists of the fact that whatever the curvature of space in the universe may turn out to be, we already know that space is *nearly* flat, neither positively nor negatively curved. An exact "flatness" of space arises if and only if the total density of the universe exactly equals the critical density. Two decades ago, when the inflationary model burst upon the scene, cosmologists summarized this requirement by stating that in order for space to be flat, Ω_M must equal 1. If the cosmological constant equals 0, as nearly everyone then believed, this statement is quite correct, but we know, as cosmologists already did, that the true and more general criterion for flat space is that Ω_M plus Ω_Λ must equal 1. A nonzero cosmological constant, which produces energy in every cubic centimeter of empty space, also contributes to the total curvature of space in the cosmos, because, as Einstein first noted, any amount of energy E creates a corresponding amount of mass equal to E/c^2, where c is the speed of light.

The flatness problem arises when we pause to consider that the sum of Ω_M and Ω_Λ could, in principle, equal any value that we might imagine, such as 0.000000000584, 7,566,898,043.732, or numbers that are even farther from 1 than these are. Each of these values for the sum of Ω_M and Ω_Λ corresponds to an imaginable universe. Even worse, unless the sum of Ω_M and Ω_Λ equals 1 at a time soon after the big bang, the expansion of the universe will quickly drive this sum to a value far from 1. In 1980, astronomers already knew that Ω_M must be greater than 0.1 and that Ω_M plus Ω_Λ cannot be larger than about 2. The minimum value of Ω_M arose from astronomers' observations concerning the density of matter, and the maximum limit on Ω_M plus Ω_Λ came from extending the Hubble diagram to immense distances: If the universe has a value of Ω_M plus Ω_Λ larger than about 2, astronomers would have found it by comparing distant galaxies' redshifts with their estimated distances.

In comparison with the immense range (to be bold, an infinite one) of theoretical possibilities for the sum of Ω_M and Ω_Λ, most cosmologists found it a startling fact, highly laden with import, that Ω_M plus Ω_Λ now has a value relatively close to 1, whether or not this sum exactly equals 1. We may view this result in another way by noting that if space in the universe is curved, it does not curve by much, considering how it would curve if the sum of Ω_M and Ω_Λ had a value far from unity, either many times larger than 1 or much closer to 0 than to 1. Though the public remained unperturbed by the flatness problem, most cosmologists took it seriously, coupling it in their minds with a second paradox about the universe.

THE HORIZON PROBLEM

The standard big-bang model of the universe, in which the cosmological constant equals zero, envisions an expanding cosmos whose rate of expansion has always been decreasing. This decrease arises quite naturally as the result of mutual gravitational attraction among all its constituents that possess nonzero masses. At any time in the history of the universe, an observer can see out to a "horizon," the distance to which equals the speed of light times the age of the universe. As time goes on, additional matter comes within this horizon, which expands at the speed of light, more rapidly than any galaxy recedes. This all makes good sense, and fits together quite rationally. Yet, the standard big-bang model leads to a result so difficult to believe that the temptation arises either to ignore it or to accept it as completely obvious.

That astonishing result consists of the fact that *the universe looks about the same in all directions.* Astronomers find approximately equal numbers of galaxies in each large volume, and those galaxies fall into similar categories whose members resemble one another no matter where they appear. Furthermore, the cosmic background radiation, a faint glow that was produced throughout the universe early in its history, arrives from all directions with nearly the same intensity and with a single spectrum that characterizes this radiation from all directions. Although the tiny deviations from this rule carry potentially immense amounts of information (as we shall see in Chapter 11), we should not lose sight of the more fundamental fact that to a high degree of accuracy, the background radiation shines upon us equally from all sides.

Why does this uniformity in all directions pose a paradox? When we look in opposite directions as far as possible, receiving light from the most distant galaxies and radio waves from the even more distant source of the cosmic background radiation (the long-vanished early years of the universe, seen at distances close to 14 billion light-years), we are observing regions of the universe that have never had causal contact, because these regions lie, and have always lain, outside each other's horizon, completely unaware of each other's existence. This means that whatever physical conditions may exist or may have existed in these regions, they have never had a chance to share information about the conditions within them at any time throughout the full history of the universe. This sharing can occur, for example, if particles and radiation interpenetrate the two regions, passing from one to the other and effectively telling each region what has been happening in the other. On Earth, for example, causal contact creates a fairly even distribution of pollutants throughout the atmosphere and oceans, as molecules collide and spread out from the sources of pollution. If this did not occur, we might expect, for instance, that the ocean near Tahiti would remain far cleaner than the ocean near Indonesia, instead of (as is unfortunately the case) the difference in the density of pollutants amounting to only a modest one.

If different regions of the visible universe have never been in causal contact, then the fact that they look almost exactly alike, so far as galaxies and the cosmic background radiation are concerned, poses a cosmological mystery. To explain why we see almost identical conditions in all directions therefore requires us to assume that the early universe had almost identical conditions throughout, so that all regions produced identical amounts of the cosmic background radiation and formed galaxies in the same types and numbers, even though each of these regions evolved essentially on its own, free of any homogenizing influence from the other parts of the cosmos. If the near identity of conditions held true to 1 part in 10, or 1 part in 100, this might be explained, or at least believed, as the result of all regions receiving a similar start after the big bang; but when the cosmic background radiation turns out to arrive (as it does) in the same amounts from all directions to an accuracy of 3 parts in 100,000, coincidence fails as an appealing explanation.

THE INFLATIONARY MODEL
EXPLAINS EVERYTHING

Cosmological thought, embodied in the finest minds of the late 1970s, stood ready for a theory of the universe that could somehow preserve the essence of the standard big-bang model while resolving the flatness and horizon problems. Intriguingly, this motivational factor hardly entered the minds of those who originally formulated the inflationary model. Among the cosmologists who had the greatest effect in creating this model, we may name Alan

Guth (then at Stanford University and now a professor at the Massachusetts Institute of Technology), Paul Steinhardt (then at the University of Pennsylvania and now at Princeton University), Andreas Albrecht (then at the University of Pennsylvania and now at the University of California at Davis), and Andre Linde (then in Moscow but now at Stanford University). These theorists and their colleagues mainly sought to explore the consequences that arose from certain theories of how elementary particles interact, when they applied those theories to the universe's earliest moments. Undaunted by the apparent nuttiness of their ideas (a judgment that, after all, would rule most of cosmology, including the standard big-bang model, right off the board), they noted that a particular class of elementary-particle theories, if true, made a startling prediction about an era quite early in the universe's history. According to these theories, during this era the cosmos must have expanded at a truly extraordinary rate, one that may leave the reader breathless even after learning that great cosmologists have approved it. Partly because of the economic troubles of the early 1980s, theorists living in capitalist countries named this epoch of rapid expansion the "inflationary era" and called their model the "inflationary universe."

How much inflation occurred during the inflationary era? According to the models that might explain why the universe presents the picture we see now, inflation involved simply enormous increases in the size of the universe. Since the universe may be infinite in size, we must refrain from naming any actual sizes; instead, we may refer to how the distance between any two representative points has changed. For example, the inverse of the present Hubble constant, $1/H_o$, would equal the age of the universe since the big bang if galaxies' recession velocities had not changed with time. This time interval also approximates the amount of time in which the distance between two representative regions—two large clusters of galaxies, for example—will double at the present rate of expansion. Since $1/H_o$ equals about 15 billion years, we may rightly say that each doubling in size of the cosmos takes a long time. During the inflationary era, however, this sort of doubling proceeded much more quickly, requiring only about 10^{-33} second, rather than the 500 million billion seconds of $1/H_o$ today. This difference corresponds to a factor of nearly 10^{51}, a number too large to be usefully expressed with the million billion trillion quadrillions of ordinary speech.

Events proceeded rapidly during the inflationary era, which lasted from perhaps 10^{-33} second A.B.B. (after the big bang) to 10^{-30} second A.B.B. In those brief moments, the universe doubled in size not a few times, but perhaps several hundred times. Let us now be brutally frank: The fact that so many doublings in size occurred within such a short span of time means that different parts of the universe were moving away from one another at speeds much greater than the speed of light. This appears to violate the principle laid down in Einstein's special theory of relativity that no velocity can exceed the speed of light; but as we shall soon discuss, no laws are being broken here.

The 200 doublings, more or less, that took place during the inflationary era increased the size of the universe by a factor of 2^{200}. Every 10 doublings increased the universe's size by 2^{10}, which equals 1,024, and this increase by a factor greater than 1,000 occurred perhaps 20 times. The net result of 200 doublings therefore approximately equals 1,000 (in actuality, 1,024) raised to the twentieth power. Since 1,000 is 10^3, this factor is 10^3 raised to the twentieth power, which equals a factor of 10^{60}.

When even a tiny volume expands by a factor of 10^{60}, it produces an enormously large one. If, for instance, a region 10^{-26} centimeter across expands by this factor, it becomes a region that spans 10^{34} centimeters. In other words, inflation can turn a region thirteen orders of magnitude smaller than a proton into a volume a million times larger than the visible universe today! The exact—or even the approximate—sizes of the small and large volumes scarcely matter. What counts is that the inflationary era could take any incredibly small volume of space and—in a mere 10^{-30} second—make it a region far, far larger than the visible universe.

What could induce this sort of immensely rapid, enormously successful increase in volume? The answer lies in the theories of particle physics that cosmologists have applied to the early universe to produce the inflationary model. These theories lie beyond the horizon of this book, except for an intriguing summary of their import. Particle physics proposes that *during the inflationary era, the universe acquired an enormous cosmological constant, which faded to zero as the universe became 10^{-30} second old.* The universe's expansion during the inflationary era exactly models—on a far more rapid timescale—the acceleration of the expansion that a nonzero cosmological constant can induce today. The inflationary era, during which the universe doubled in size repeatedly, corresponds to the de Sitter model of a cosmos without matter, which increases in size exponentially. During that era, matter in the conventional sense did not exist in the universe; instead, a tremendous density of energy, embodied in a cosmological constant, drove the inflation and evolved into matter plus energy as the inflationary era ended. (Recall that the energy brought into the universe by the existence of a nonzero cosmological constant amounts to an effective density, because energy always implies an equivalent amount of matter, as summarized by Einstein's $E = mc^2$.)

INFLATION EXPLAINS THE FLATNESS AND HORIZON PROBLEMS

Thus the inflationary model of the universe amounts to the assumption, grounded in modern theories of how elementary particles interact, that a tiny region of space, soon after the big bang, could acquire an enormous but transitory "cosmological constant." During the 10^{-30} second or so that this constant differed from zero, it would turn this tiny region into one larger than we can easily imagine. After the cosmological constant had fallen to zero, the universe would expand according to the standard big-bang model, just as it would have done if the cosmological constant had always been zero.[1] An observer adrift in the post-inflationary universe would observe the same conditions as an observer in a cosmos that had never undergone inflation—except that the horizon and flatness problems would be solved.

Solved—how? By the fact that the tiny region that inflated did maintain causal contact throughout its volume and by the equally agreeable fact that whenever a region of space expands by such an enormous factor as 10^{60}, it inevitably becomes flat. Let us take these two facts in turn, as they help us to celebrate the triumph of the inflationary model as a theoretically reasonable explanation of two great problems confronting modern cosmology.

The horizon problem deals with the issue of causal contact, that is, of different regions of space being able to share information. This sharing occurs most straightforwardly by the interpenetration of matter and radiation between the regions. Our earlier reasoning leads us to the conclusion that in the standard big-bang model, in

---

[1]The statement that the cosmological constant fell to zero actually means that the constant decreased to the relatively tiny value that may exist today.

which no inflation occurs, two regions of space can establish causal contact only if their separation does not exceed the age of the universe, T, multiplied by the speed of light, denoted by c and equal to 3×10^{10} centimeters per second. The product of c times T equals the distance that light and other forms of electromagnetic radiation have had a chance to travel at a time T after the big bang. At a time 10^{-33} second after the big bang, for example, the largest volume that could have maintained the same physical conditions throughout spanned a distance equal to 10^{-33} second multiplied by 3×10^{10} centimeters per second, or 3×10^{-23} centimeter.

Now suppose that a region much smaller than this—for example, one 10^{-24} centimeter across—within which uniform conditions did exist happened to expand much more rapidly than the speed of light. This inflation, as we have learned to call it, would effectively spread the region's uniform conditions into a prodigious volume. When the inflationary era had ended, the once-tiny region would have grown 10^{60} times larger. The density of matter and radiation within the region would therefore have decreased by an enormous factor, but these parameters, along with all others describing the physical situation within the inflated volume, would have fallen by precisely the same factor throughout the now-immense region. As a result, the inflationary era would have produced a colossal region of space with essentially identical conditions throughout its volume.

Of course, we must pay a price for believing this: The region of space that turns out so well must expand far more rapidly than the speed of light during the inflationary epoch. How can this be possible? Doesn't Einstein's theory of relativity forbid any motion at speeds greater than the speed of light?

Not completely, comes the answer from the physicists. Careful examination of Einstein's special theory of relativity—an examination that began as soon as Einstein published it and has continued to the present—shows that the theory forbids only *local* motions that exceed the speed of light. The word "local" here refers to objects that occupy the same vicinity and pass by one another at relatively modest separations. Relativity theory prohibits a satellite from orbiting the Earth at speeds greater than light speed or an astronaut from leaving the solar system at a velocity greater than c. But the theory does not bar distant parts of the universe from receding at speeds greater than the speed of light. This important distinction, which may seem a bit weaselly but in fact helps to underscore what the special theory of relativity actually describes, allows the universe to expand more rapidly than the speed of light. It turns out that even in the standard big-bang model of the universe, different regions move apart from one another more rapidly than the speed of light. Even so, the standard big-bang model cannot really explain the horizon problem.

The inflationary model, in contrast, produces a natural explanation: What we call the "universe," the region of space surrounding the Milky Way and far larger than the limits of our observation, once expanded far more rapidly than the speed of light. After this inflation had blown a tiny bubble of space into an enormous volume, it left that volume with identical physical conditions throughout. The billions of years of evolution and expansion after the inflationary era have changed these conditions tremendously, but the visible universe continues to maintain, and to display, nearly the same state of affairs within any volume that provides a good sample of its contents.

So much for the horizon problem, which turns out to have an easy explanation once we abandon our hope and belief that the cosmic speed limit set by the speed of light could be truly universal. What about the flatness problem? This, too, melts away, reaching a harmonious accord with our understanding as the result of inflation. The flatness problem refers to the fact that we find the sum of Ω_M and Ω_Λ far too close to unity to be acceptably explained as the result of chance. If this sum exactly equals 1, then space must be perfectly flat, neither positively curved, like the surface of a sphere, nor negatively curved, in analogy with the two-dimensional surface of a saddle.

But the inflationary theory predicts that space should be perfectly flat! No matter what the curvature of space may have been before the inflationary era began, the increase in size by a factor of 10^{60} or so would inevitably have made space effectively flat. More precisely, inflation would have made any small region of space, such as the visible universe today, seem almost perfectly flat (to about 1 part in 10^{60}), just as a tiny fraction of a balloon's surface seems flat to those who remain within that region. If inflation did in fact increase the size of a once-tiny region of space by a factor of 10^{60}, then, as we have seen, the entire visible universe, extending 15 billion light-years in all directions from us, would span less than 1 part in 10^6 of the inflated volume. In that case, everything we can observe, or can hope to observe, within the volume that we call the visible universe amounts to looking at less than one square millimeter on the surface of a balloon the size of a town. Not surprisingly, this square millimeter would seem almost perfectly flat. The inflationary theory thus explains the flatness problem through our limited horizons: We perceive the universe to be nearly flat because all the space that we can ever see *is*, in fact, flat.

THE CRUCIAL PREDICTION: IF INFLATION IS CORRECT, SPACE MUST BE COMPLETELY FLAT—OR MUST IT?

When the inflationary theory first burst upon the scene, in the early 1980s, all astronomers save a hardy bunch of cosmological theorists found it more amusing than satisfying. The mental creation of an inflationary era, during which the size of the universe increased by something like a factor of 10^{60}, seemed stretching a bit too much to provide an explanation for the horizon and flatness problems. (The inflationary theory also explains another issue, called the "magnetic monopole problem," but even this success may not appear to justify the amazing assumptions embodied in the theory.) Cosmologists received a fair amount of good-natured ribbing (is there any other kind in cosmology?) over the lengths to which they would extend themselves in favor of the inflationary model of the universe, without any evidence for inflation except the convoluted argument about the two problems we have examined.

Defending themselves nimbly (an activity at which every theorist must excel, or otherwise choose another line of work), the pro-inflation lobby pointed out that the inflationary model makes a definite prediction, one that could before long be tested with improved observations of the cosmos. This argument did indeed impress the doubters, as it ought to have done: Scientists despise theories that make no testable predictions, admire those that do make such predictions, and love those whose predictions prove correct upon further testing. Since the early 1980s, the inflationary model's prediction that Ω_M plus Ω_Λ equals 1 has served as a beacon and a guide for cosmological theorists. The compelling power of the model appears in the fact that even during the decade following the early 1980s, when astronomers' best estimates set Ω_M at about 0.1 or 0.2 and maintained Ω_Λ at a flat 0, many theorists remained convinced that the sum of the crucial parameters Ω_M and

Ω_Λ must nevertheless equal unity! From the theoreticians' standpoint, it was the laggard observational astronomers who were failing in their proper task: to discover either an additional density of matter (expected to occur by most theorists) or a nonzero cosmological constant (favored by a few), which would allow observational data to agree with theory, by demonstrating that Ω_M and Ω_Λ indeed sum to unity.

During the mid-1990s, when the sum of Ω_M and Ω_Λ seemed to fall well short of 1, cosmologists demonstrated their supple minds by creating "open-inflation" models that allowed for this possibility. To do so, they envisioned an inflationary epoch in which a smaller number of doublings occurred than in the "standard" inflationary model. In that case, the immense expansion of the cosmos during the inflationary era would leave the cosmos nearly, but not exactly, flat, as if we lived on a balloon that had expanded from a submicroscopic bubble to "only" the size of the Earth, which we can determine to be not quite flat. The open-inflation models produce a universe in which the sum of Ω_M and Ω_Λ equals any number one might like. To those on the outside, who had admired the basic inflationary model for making a clear, definitive prediction that astronomers could test by determining the curvature of space, the advent of open-inflation models removed much of the beauty of the concept. "Desperate measures by desperate people," commented Paul Steinhardt, who favors the standard model of inflation. To be sure, beauty or the lack of beauty does not determine the truth, whatever John Keats may have thought. The first order of business remains the determination of the key parameters that describe the cosmos, which will match some theoretical models while rejecting others.

The exciting cosmological news from 1998 implies that the original inflationary theory appears to make the correct prediction: Ω_M and Ω_Λ *do* sum to a value close to 1. Having achieved a thorough understanding of theoretical cosmology, we may proceed to examine the current state of cosmological observations. We shall first consider the dark matter in the universe, which furnishes most of the total value of Ω_M, and shall then turn to the stunning news from the cosmological front concerning the value of Ω_Λ.

CHAPTER SIX

DARK MATTER RULES

FOR CENTURIES, ASTRONOMERS have been attempting to make a complete census of the matter in the universe. Long before cosmological theory revealed that the average density of matter could determine the destiny of the universal expansion, a natural curiosity had led those who study the cosmos to record what they saw and to calculate the total amount of matter per unit volume. Although the task of cataloging the amounts of mass that appear in different types of cosmic objects may seem routine and even boring, this avenue of investigation has proven a broad highway to a new understanding of the universe. One of the greatest astronomical discoveries of the twentieth century deals with the slowly accepted, but now undeniable, realization that the universe contains far more mass in an unknown form than anyone can account for with the types of particles we do know. In other words, in terms of mass, most of the universe remains a complete mystery, its existence revealed to us only by its gravitational effects on the mass that we do see and can understand.

When members of the public look beyond the limiting horizon of "space"—past the human and automated astronauts we have sent to explore our tiny corner of a subsection of the nearby volume of the Milky Way—they encounter a gallery of fantastic objects, enough to satisfy almost any desire to encounter objects that rank as strange beyond imagination. Quasars, pulsars, and blazars; white dwarfs, neutron stars, and black holes; exploding stars and gamma-ray bursters; star-forming nebulosities and dusty, pancake-shaped clouds that give birth to new planets—

these residents in the cosmic zoo seem quite sufficient, along with the strange behavior of space itself, to consume all the mental and emotional energy that we can bring to bear on the universe.

Yet we must accept that the totality of all these objects ranks as a minority fraction of all the matter in the universe. The cosmos consists mainly of other stuff, about the nature of which we can say next to nothing. That this unknown matter dominates the gravitational effects that rule the universe, and therefore holds the fate of the universe in its ill-described hands, seems barely plausible. But so it is.

HOW TO FIND DARK MATTER
THROUGHOUT THE UNIVERSE

Astronomers introduced the phrase "dark matter" to describe what they cannot see but nevertheless can detect. To astronomers, this name connotes not simply that the matter emits no light, as the word "dark" conventionally implies, but also that dark matter produces no electromagnetic radiation at all, so it remains invisible over the entire electromagnetic spectrum, from the longest-wavelength radio waves to the shortest-wavelength gamma rays. Though astronomers have opened one spectral window after another with new satellite observatories sent above the absorbing effects of our atmosphere, most of the dark matter, as we shall see, consists of matter that cannot be directly detected by these or any other improved instruments of the future.

Not surprisingly, astronomers have been slower to find dark matter than to observe matter that does emit some type of electromagnetic radiation. Operationally, we can imagine two entirely different techniques to search for dark matter in the universe. First, we may try to find dark matter right here on Earth, capturing some of the dark matter, even though we do not know its nature, as it passes by or through our planet. The search for dark matter with this approach, barely begun and so far unsuccessful, lies at the frontiers of particle physics. The second method detects dark matter by observing the results of its interactions with matter that we do see. This method has two subcategories. In one, which can detect only part of the dark matter, astronomers examine the relics of nuclear-fusion processes in which the dark-matter particles have participated. In the other, which has yielded the most significant results, they detect dark matter by observing its gravitational effects on the visible matter. So long as the dark matter has mass (which it must have if we are to call it "matter"), it exerts gravitational forces on all other objects with mass in the cosmos. This fact, pursued in practice by the expert astronomers who specialize in observing galaxies and galaxy clusters, has led to the revolution in our understanding summarized by the pithy statement that most of the cosmos consists of dark matter of unknown composition.

DETECTING DARK MATTER BY
ITS GRAVITATIONAL EFFECTS ON VISIBLE MATTER

Astronomers first detected dark matter by observing the motions of stars in galaxies and of galaxies in galaxy clusters. Those who know astronomical history may believe this statement to be incorrect; they recall that well over a century ago, astronomers had shown that dust grains floating among the stars in the Milky Way concentrate heavily in certain regions, where the density of dust rises to the point that it blocks the passage of starlight, creating dark lanes in the Milky Way that we see on the sky. Surely this dust qualifies as dark matter! But it does not. Astronomers can now detect infrared emission from the closest regions containing this interstellar dust, because the dust grains, though too cold to gleam in visible light, have sufficiently high temperatures to produce longer-wavelength infrared radiation.

True dark matter first made its presence known when astronomers used the Doppler effect to measure the velocities of galaxies in galaxy clusters. These speeds depend on the strength of the galaxies' mutual gravitational attraction: Larger attractive forces provoke more rapid motion. Many of the galaxies in clusters seemed to be moving more rapidly than astronomers could explain on the assumption that a galaxy's mass resides mainly in its stars, whose individual masses were known and whose total numbers could be well estimated from the galaxy's apparent brightness. In 1933, Fritz Zwicky, an intriguing maverick scientist whom we shall meet again, stated his conclusion that "in many of the largest galaxies, the amount of dark matter is comparable with that of the visible matter." Zwicky's bold conclusion, which seemed to assign dark matter an importance yet to be verified, has proven entirely modest, now that astronomers know, or think they know, the total contribution that dark matter makes to the universe.

Astronomers could and did employ the technique of looking for dark matter by its gravitational effects within individual galaxies, most notably within our Milky Way. Like galaxies in a galaxy cluster, stars in a galaxy move in response to the total gravitational force that acts upon them. In measuring the motions of individual stars, astronomers observe these motions with reference to the center of the galaxy. This technique reveals the amount of mass within the galaxy itself, rather than the masses of other galaxies, because the other galaxies in the universe, massive though they are, lie so far away that they attract all the stars in the galaxy with nearly the same amount of force and do not significantly affect how a star moves with respect to the center of its own galaxy.

A star's motion therefore reveals the combined gravitational effects acting upon it, produced by all the objects with mass in its galaxy. Any hope of using this fact to deduce how much matter occupies different parts of the galaxy might seem to require superhuman ability: Astronomers must attempt to discriminate among the effects that different regions produce on a particular star, even though they see only the total effect produced by all of them. Fortunately, Isaac Newton discovered a little trick that works wonders in this effort. (Newton was actually concerned with the combined gravitational effects that different parts of the Earth produce on an object, but his genius was such that his discovery has a universal application.) Imagine a star in a galaxy such as our Milky Way, moving in a nearly circular orbit around the center of the galaxy. Although the star feels the effects of the gravitational forces from the hundreds of billions of other stars in the galaxy, from all the interstellar gas and dust, and from all the dark matter that the galaxy contains, we can mentally divide everything that exerts gravitational force on the star into two regions: one that contains everything that lies closer to the center of the galaxy than the star; the other, the volume that lies *farther* from the galaxy's center.

Newton showed that if we make the reasonable assumption that the mass in the galaxy has a symmetric distribution in the different directions outward from the center of the galaxy, the situation simplifies itself remarkably. The gravitational forces from all the matter farther from the center *cancel* one another, so that their net gravitational effect equals precisely zero. Thus the star's motion does not depend on either the amount or the overall distribution of the matter more distant from the galactic center. As for the matter that lies closer to the center, an equally marvelous simplification appears: All the matter closer to the center exerts a total gravitational force on the star precisely equal to what would occur if a single object occupied that center, the mass of which amounted to the total mass of everything closer to the center than the star under examination.

In short, matter farther from the center doesn't count, whereas matter closer to the center amounts to the equivalent of a single object at the center with a mass equal to all the closer-in mass. Suddenly the hundreds of billions of stars, the interstellar dust, and the unknown number of dark-matter clumps or particles reduce in effect to a single object, as simple in gravitational terms as the sun that lies at the center of the solar system. Like a single planet orbiting a star such as our sun, the representative star in a galaxy will orbit around the center of the galaxy in response to the combined gravitational force from all the matter closer to the center.[1]

Astronomers can therefore attempt to "weigh" a galaxy by measuring the speeds at which stars move in orbit around its center. The velocity of each group of stars at a particular distance from the center reveals the corresponding amount of mass that lies closer to the center. By using the Doppler effect to measure the velocities of groups of stars in the Milky Way or in a nearby galaxy, astronomers can deduce the total amount of mass, visible or dark, within a nested set of distances from the galaxy's center. During the 1970s, the astronomer Vera Rubin and her collaborators used this method to obtain a startling result about giant galaxies, including our own. Although stars appear to provide most of the mass within the volume that provides most of the galaxy's light, immense amounts of dark matter lie far beyond these regions. Of the galaxy's total mass, less than one-tenth resides in stars and more than nine-tenths in these outer parts replete with dark matter! A giant galaxy such as the Milky Way may contain several hundred billion stars, the mass of which totals 200 or 300 billion times the sun's mass, but its total mass can surpass 5 trillion solar masses. Nearly all of this mass consists of dark matter, most of which lies farther from the center than all but the most distant of the galaxy's stars.

---

[1] This statement holds strictly true only if the matter in the galaxy has a spherically symmetric distribution, but in the actual situation, it provides a close approximation to reality.

The results from the Milky Way and nearby galaxies, combined with observations of the motions of more distant galaxies that belong to large clusters of galaxies, have achieved a reassuring congruence that places dark matter's domination of the gravitational universe beyond reasonable doubt. Giant galaxies contain at least ten times, and possibly as much as twenty to forty times, more dark than visible matter.

How does this affect the quest to find the value of Ω_M, the total density of matter in the universe? Astronomers' current best estimates set the contribution to Ω_M of the matter that resides in stars at about 0.005. In other words, visible matter in the universe has a density equal to one-half of 1 percent of the critical value. This result depends on the exact value of the Hubble constant, which astronomers derive from the distances to galaxies that they have evaluated. Because the value of the Hubble constant has an uncertainty of 10 or 20 percent, the measured average density of the visible matter, which depends on the square of the value of the Hubble constant, remains uncertain by 20 to 40 percent.

Seen from the perspective of whether Ω_M equals 1, this hardly seems relevant. Visible matter does not provide even 1 percent of the critical value of the density: Stars by themselves fall hopelessly short of creating a flat universe or of bringing the expansion to a halt if the cosmological constant equals 0. But when astronomers found that the amount of dark matter equals some fifty times the amount of visible matter, these differences acquired significant implications. The current best estimates for the amount of dark matter rate its total contribution to Ω_M at 0.2 to 0.3. This value implies that the dark matter so completely dominates the matter in stars that we may neglect the stars in assessing the future of the universe. Straining to the upper bound of currently allowable errors of measurement, some astronomers will concede that the average density of dark matter may amount to 40 or possibly even 50 percent of the critical value, so that Ω_M might be as large as 0.4 or 0.5.

Cosmologists agree, however, that unless we discover a significant error in astronomical observations, dark matter cannot take Ω_M all the way to unity. Though dark matter has raised the upper bound of the deduced average density of matter to about 40 percent of the critical value, the final factor of 2.5 remains an insurmountable barrier. Before we set the stamp of finality on this conclusion, we should examine another method that astronomers can use to see how matter's gravitational effects relate to the average density of matter: the theories that explain how galaxies formed.

GALAXY FORMATION AND
THE TOTAL DENSITY OF MATTER

Thirteen billion years ago, give or take a billion years or two, the almost featureless and formless universe began to generate galaxies. Today we see a cosmos full of galaxies, the "island universes" that Harlow Shapley once pooh-poohed but that contain nearly all the visible matter in the depths of space. One of the great frontiers of astronomy—a mystery par excellence that everyone knows can and must be solved—deals with the formation of these galaxies, including the immense amounts of dark matter that dominate the galaxies' outer "halos." The epoch of galaxy formation, which occupied the first one to three billion years after the big bang, provides us with one of the most important and least understood eras in the history of the universe: the dark ages when no stars shone in galaxies.

Current theories of galaxy formation assume that galaxies have grown from modest fluctuations in the density of matter that already existed a million years or so after the big bang. Studies of the cosmic background radiation, which we shall examine in Chapter 11, have begun to reveal these fluctuations to our direct observation. This success has given birth to a cottage industry of cosmologists who attempt to make computer models of how the first tiny variations in density eventually produced much denser clumps of matter and how these clumps continued to contract and to grow denser until they turned into galaxies, made of matter at a density sufficient to induce the formation of stars.

Generations of once-young astronomers have employed generations of improved computers in their attempts to model the early universe as it changed from a nearly featureless froth into the extraordinarily complex arrangement it displays today, with many different types of galaxies and countless detailed variations on the basic types. Everyone knows that the answer to the mystery of galaxy formation lies with the attractive force of gravity, the force expressed by all matter as an attraction for all other matter. The trick is to model the formation of galaxies through gravitational forces, starting with tiny deviations in density and arriving at today's universe, in which enormous clumps of matter—galaxies and galaxy clusters—have an average density hundreds of thousands of times greater than the density of matter between the clusters.

Among their other aspects, the computer models differ in the densities they assume to have existed as galaxies formed. This density, which declined continuously as the universe expanded, has a direct correspondence with the total average density of matter today. Hence, if cosmologists can determine which model of galaxy formation best matches the universe we see around us, they will have a reliable, if roundabout, estimate of Ω_M, the current average density of all matter. Furthermore, this value will refer to the entire universe, not simply to individual clumps of matter such as galaxies and galaxy clusters. Astronomers cannot be certain that a galaxy cluster provides them with a representative volume of space, so far as the average density of matter is concerned; the dark matter, like the visible matter, might well have clumped together, leaving a lower average density between galaxy clusters than within them. The determination of Ω_M from models of galaxy formation offers a more universal result than finding Ω_M by detecting the dark matter in limited, though enormous, regions of space.

Unfortunately, cosmologists have not yet reached a definite determination of Ω_M by comparing computer models with observational data. Their current best estimates of Ω_M using this technique do set its value close to 0.3, not far from the value of Ω_M found for galaxy clusters, which seems to lie closer to 0.2. Because this method of modeling the formation of galaxies plays such an important role in estimating the value of Ω_M, we shall return to it for a closer look in Chapter 12. For now, we have larger fish to fry: We must accustom ourselves to discriminating among different types of the dark matter that we cannot see.

THE THREE TYPES OF DARK MATTER

Gravitational forces, which arise from all forms of matter, can be deduced in the present by observing motions within galaxies and galaxy clusters, and they can be deduced in the past by computer models of galaxy formation. These techniques reveal that the great bulk of all matter takes the form of dark matter, which produces no electromagnetic radiation for astronomers to analyze. The pellucid depths of cosmological insight receive an excellent encomium from the fact that *even though we have no idea what constitutes the dark matter, we must divide it into at least three separate categories if we hope to make good sense of the universe.* These categories are (1) hot dark matter, (2) cold dark matter, and (3) ordinary dark matter, also called "baryonic dark matter." Let us examine these in reverse order to see why we must create seemingly complex categories to deal with what seems a single unknown quantity.

Baryonic or ordinary dark matter includes all the dark matter that participates in nuclear-fusing reactions. Physicists use the term "baryons," which means "heavy particles," to denote the particles that undergo nuclear fusion. Of all baryons, the protons and neutrons that compose all the nuclei of atoms are by far the most familiar to us. When cosmologists use the term "baryonic matter," they do not mean to exclude other familiar particles, such as electrons, that do not fuse together as nuclei do. Instead, they mean simply to describe the matter that we already know, more accurately described as ordinary matter. In other words, baryonic dark matter means dark matter made from any particles already known to science. This baryonic dark matter might consist of individual nuclei, of dust grains, of rocks, or of satellite-sized objects. The similarity among these forms—the fact that they all consist of ordinary matter—overshadows their apparent differences, which simply reflect how effectively nuclei have clumped together.

To aid in their considerations of the universe, cosmologists divide all possible types of nonbaryonic or exotic dark matter into two categories, summarized as hot and cold. These are cosmological shorthand terms that describe the average speed of the (unknown) types of particles that constitute the nonbaryonic dark matter. If these particles have masses comparable to those of a proton or neutron, then at the time when matter first had the opportunity to clump together, its constituent particles, like the familiar nuclei, would have been moving at speeds much less than the speed of light. By cosmological definition, this matter qualifies as "cold."

Particles with much smaller masses will have been moving at nearly the speed of light at the time when clumping first became possible, 300,000 years after the big bang. Cosmologists describe these particles as "hot." Hot dark matter includes neutrinos, a type of particle that we already know, which once were thought to have zero mass but which now appear to have tiny but nonzero masses. As a matter of complex fact, neutrinos come in three varieties, of which only at least two, according to our current knowledge, definitely have nonzero masses. This merely underscores the messy nature of "hot dark matter," which cosmologists use as a catchall phrase to include both hot dark matter of unknown form and the hot dark matter that consists of neutrinos with nonzero masses.

The reason for discriminating between cold dark matter and hot dark matter lies in the fact that hot and cold particles behave quite differently as the cosmos attempts to form clumps. Cold dark matter, whatever the details of its composition, clumps together far more readily, because the comparatively low velocity of its constituent particles permits gravitational forces to bring them together with relative ease. Attempting to make clumps of hot-dark-matter particles, in contrast, resembles shoveling fleas across a barnyard: Their tendency to escape overwhelms gravity's ability to produce and to maintain a coherent clump of matter.

In Chapter 12, we shall see how the distinction between hot-dark-matter and cold-dark-matter particles enters cosmologists' attempts to model the formation of galaxies. For now, we may pause to note that we have divided all matter in the universe into four categories:

1. Ordinary matter, the sort that we know, participates in nuclear-fusion reactions and is referred to by physicists as "baryonic." We may think of this as the cosmic "earth," the familiar grit that forms animals, planets, stars, and galaxies.
2. Hot dark matter consists of dark-matter particles with tiny masses, which therefore had speeds close to the speed of light in the era when galaxy formation began. We may depict hot dark matter as the "air" of the cosmos.
3. Cold dark matter's name hides the fact that it refers only to non-baryonic cold dark matter, leaving all baryonic matter as a separate component. This portion of the cosmos, which almost certainly amounts to the dominant one in terms of mass and density, we shall envision as the "water" of the universe to honor its slippery, hard-to-grasp nature.
4. Photons form the universe's cosmic "fire." Most of these photons appeared immediately after the big bang, as the cosmic fireball

began its expansion, but many others were created later in stars, stellar explosions, and more exotic sources.

Earth, air, water, and fire; baryonic matter, hot dark matter, cold dark matter, and photons. These categories do not include everything that we may encounter in the universe, but they come sufficiently close for most cosmological purposes. Consider, for example, gravity waves, first postulated by Einstein and now demonstrated to exist by the exact correlation between Einstein's theories and observations of what happens when objects orbit around their common center of mass. Gravity waves cannot be called matter of any kind; nor are they electromagnetic radiation, which consists of photons only.

Another cosmic substance deserves immediate attention because it affects the future of the universe: the cosmological constant, a mysterious entity that amounts to an energy that appears throughout empty space. We shall examine the nature of this beast in greater detail in the next chapter. For now, we should not be surprised to learn that some of the cosmologists who speculate about the cosmological constant have resurrected the name "quintessence," the fifth essence of the ancient and medieval cosmos, thought to be a perfectly pure substance, quite unlike the fire, air, earth, and water that constitute our terrestrial realm. Ancient Greek astronomers believed that the quintessence forms the sun, moon, planets, and stars, as well as the transparent spherical shells then believed to rotate around the Earth, carrying all celestial objects with them. Today no one imagines that the Earth consists of material that differs fundamentally from the rest of the universe or that we can describe the cosmos in terms of just five entities. Nevertheless, to keep pace with the discoveries of modern cosmology, we may chant the mnemonic mantra: Earth signifies the baryons, air stands for the hot dark matter, water implies the cold dark matter, fire deals with the photons, and quintessence represents the cosmological constant, which is not matter at all, but an energy that lurks in empty space.[2]

---

[2]Theoretical cosmologists have actually assigned the name "quintessence" to the cosmological constant only in those model universes in which the size of the "constant" changes with time. If the cosmological constant indeed has a constant value, as its historically established name implies, then cosmologists do not call it quintessence. This subtlety need not derail us long in our attempt to create a mnemonic device to remember the five constituents of the cosmos.

Big-Bang Nucleosynthesis Reveals
the Amount of Baryonic Dark Matter

On a more mundane front, we owe it to ourselves to examine a lovely approach to cosmology that allows astronomers to measure the average density of all the baryonic matter in the universe, whether or not it shines in stars. Most cosmological research either studies objects so far away that our finest instruments can scarcely perceive them or detects radiation so faint that astronomers can barely tease its properties from a sea of interfering radio noise. In contrast, this approach relies on observations that can be made with relative ease and need not involve objects at the far reaches of the universe.

Those observations consist of measuring the relative abundances of different types of atomic nuclei, relics of the era of "big-bang nucleosynthesis," the first few minutes that followed the big bang. These relics embody the history of what occurred during those first furious minutes, when the entire universe seethed with fantastic amounts of energy that could fuse nuclei anywhere in space. Now that things have quieted down because of the ongoing expansion, the nuclei that formed during that era still carry the record of what happened then. In particular, they can tell us, with a precision that may seem astounding, the density of baryonic matter throughout the era of big-bang nucleosynthesis. This density, in turn, corresponds directly to the average density of baryonic matter in the universe today, one of the crucial parameters describing the universe that cosmologists would give their eye-teeth to know.

The Record of Early Nucleosynthesis

Fully aware, then, that we are on the trail of only ordinary (baryonic) matter, we may ask, What does the record tell us about the nucleosynthesis that this matter has undergone? In order to answer this question, we must be sure that the record we examine—the relative numbers of nuclei of different types that we now find in the cosmos—arose during the era of big-bang nucleosynthesis, and not during some later era, when local conditions rather than the universe as a whole governed the creation and destruction of nuclei.

This requirement poses serious problems, because every star in the cosmos processes nuclei through nuclear-fusion reactions. We must be grateful for this fact, because nucleosynthesis in stars has created essentially all the elements heavier than hydrogen and helium, without which we could not exist, but it does introduce serious confusion into our attempts to probe the earliest minutes of the universe. The era of big-bang nucleosynthesis ended, for all practical purposes, without doing more than fusing hydrogen into helium nuclei. This reflects the reason why we can restrict this era to the first few minutes after the big bang. As the universe expanded during its first minutes, the density of baryonic matter declined continually and precipitously. This meant that the temperature likewise declined, because the particles of matter continuously found themselves with more space in which to roam. A similar effect, though not involving nuclear fusion, appears when gas is released from a spray can: The gas cools as it finds itself freed into a larger volume, a fact that appears in the cooling touch of the spray on your skin.

Because its density and temperature were falling, the early universe had only a limited amount of time to perform nuclear fusion—the time before the temperature fell below the many millions of degrees required for fusion to occur. The temperature of a gas measures the average kinetic energy per particle, so a large decline in temperature corresponds to a sizable drop in the particles' energies. Eventually, the energies all fall to the point that no collision between particles can produce nuclear fusion. Instead, the particles bounce off one another, typically repelled by their mutual electromagnetic forces, without approaching so close that fusion results.

The few minutes of big-bang nucleosynthesis began with ordinary matter mostly in the form of protons (hydrogen nuclei), which underwent wholesale fusion into helium nuclei throughout the universe. As a result, when the era of big-bang nucleosynthesis came to an end, a few minutes after the big bang, about 25 percent of the total mass of ordinary matter had taken the form of helium-4 nuclei, each of which contains two protons and two neutrons. Almost all of the remaining 75 percent of ordinary matter continued to reside in the form of protons. The first few minutes after the big bang did not provide time to produce significant amounts of any nuclei heavier than helium through additional nuclear fusion. Here "significant" refers to abundances greater than about one part in a million. The era of big-bang nucleosynthesis did manage to fuse a small amount of the total—much less than 1 percent—into rare isotopes of hydrogen and helium. These isotopes, deuterium and helium-3, deserve further attention, for they bear the record of big-bang nucleosynthesis that can tell us the average density of ordinary matter in the universe today.

Deuterium and Helium-3: The Relics We Seek to Measure

Deuterium, the isotope of hydrogen whose nuclei consist of one proton plus one neutron, also has the name "hydrogen-2," with the "2" indicating that every deuterium nucleus contains exactly two "nucleons," a word that denotes either a proton or a neutron. Long familiarity has led physicists and astronomers to call this nucleus a "deuteron," the root of which comes from the Greek word for two (which we also find in the biblical Deuteronomy). Deuterium consists of atoms that each have a deuteron as a nucleus, with a single negatively charged electron in orbit, just as in ordinary hydrogen atoms, the nucleus of which consists of a single proton and thus can be called "hydrogen-1." Once we leave hydrogen, the lightest and simplest of the elements, the numbers of nucleons in each nucleus receive more frequent citation. Helium-3, the rare isotope of helium, embraces all nuclei with two protons and one neutron. The common variety of helium, helium-4, has nuclei made of two protons and two neutrons.

During the first few minutes after the big bang, nucleosynthesis produced copious amounts of helium-4 but far less deuterium and helium-3. This occurred because deuterium and helium-3 nuclei fuse relatively easily with protons or other deuterium nuclei, creating still-heavier nuclei that contain a larger number of nucleons. In sharp contrast, helium-4 nuclei are loathe to fuse, either with themselves or with other nuclear types. Nature, which seems to love helium-4 nuclei, has ordered herself so that the nucleons in these nuclei are particularly tightly bound to one another. As a result, once nucleons find themselves within a nucleus of helium-4, they are likely to remain there.

This explains why big-bang nucleosynthesis failed to progress, in any significant way, beyond the stage of creating helium nuclei. Indeed, if two helium-4 nuclei do manage to fuse together, the resulting nucleus, one of beryllium-8, will quickly "decay" or fall apart into two more helium-4 nuclei. To make heavier elements, we need situations in which a third helium-4 nucleus will fuse with the newly made beryllium-8 before it can decay, yielding a nucleus of carbon-12. The first few minutes after the big bang did not provide this opportunity; fortunately for us, many stars do allow fusion to proceed to the point of manufacturing carbon-12 and still-heavier nuclei, such as oxygen-16, neon-20, and magnesium-24, all of which are essentially made from subunits of helium-4.

Hence we can learn little about the early minutes of the universe by studying star-made nuclei such as carbon-12 and neon-20. Instead, our hopes for probing the earliest cosmic moments with nuclear relics rests on deuterium and helium-3, which were made in large quantities during the first few minutes. But doesn't nuclear fusion inside stars also produce these nuclear types? And if so, how can we discern which of these nuclei have been unchanged since the era of big-bang nucleosynthesis?

The happy answer, much beloved by astronomers, turns out to be that on balance, stars consume deuterium and helium-3 rather than creating these nuclei. Although nuclear fusion in stars may produce some deuterium and helium-3, it invariably destroys more than it creates. As a result, when astronomers measure the amount of deuterium and helium-3 in different parts of the universe, they proceed with reasonably high confidence that they are detecting nuclei made soon after the big bang, not much later inside stars. This confidence allows them to determine the ratios of the amounts of deuterium and helium-3 to the total amounts of hydrogen and helium nuclei, and then to compare these measured ratios with those that appear in models of the cosmos that differ in the total density of baryonic matter. The measured value should match one of these models, identifying a particular density of baryonic matter as characteristic of the real universe. Deuterium has proven so useful in this approach that Michael Turner, one of the world's leading theoretical cosmologists, likes to call deuterium nuclei "the cosmic baryometer."

The baryometer's most recent results show that the average density of baryonic matter in the universe today corresponds to an Ω_M

of 0.04. This density exceeds the density of all the matter that shines in stars by about a factor of ten. Thus the cosmos presents a nested set of dark-matter problems: The baryonic dark matter, whatever form it may take, contributes ten times more to the density than baryonic visible matter does. Nevertheless, the totality of dark matter must be mainly nonbaryonic, for it possesses a total density four to ten times greater than the density provided by the baryonic dark matter! Even though baryonic dark matter far overshadows the visible matter, nonbaryonic dark matter rules over all baryonic forms, dominating the cosmos in all weighty matters that depend on gravitational forces.

If the reader has already managed to absorb the five essences of the cosmos, the nested dark-matter problem should arouse little concern. The heart of the matter still lies in attempting to determine whether Ω_M exceeds 1. Or there the heart seemed to lie—until quintessence took over where earth, air, water, and fire had failed. It is time to take the plunge into new cosmological models, resurrected once more when astronomers announced that Einstein's "greatest blunder" held the right idea after all.

CHAPTER SEVEN

SUPERNOVAE REVEAL
THE ACCELERATING
UNIVERSE

IN THE DAYS WITHOUT COSMIC ACCELERATION

For cosmology, the mid-1990s were a time of achievement, with yet-greater discoveries to come, thanks in large part to the new instruments scheduled to be sent into space at the end of the century. In those bygone days, astronomers seemed on a course toward a reasonably happy convergence in their quest to determine the key parameters that describe the universe. By 1996, improved observations suggested strongly that the Hubble constant, H, has a value that sets $1/H$ at 15 billion years, give or take a billion. If gravitational forces had never acted to slow the expansion, $1/H$ would correspond to the age of the universe since the big bang. Some deceleration has in fact occurred in the rate of expansion, so the time since the big bang must be less than $1/H$ and perhaps equal to 13 or 14 billion years. Nevertheless, the ages of the oldest stars, which astronomers estimate at 11 or 12 billion years (though some astronomers would raise this a billion or two), still appeared to allow (barely!) for the universe to have expanded for a billion years before stars and galaxies formed in earnest. Cosmologists had almost completely resolved the age problem that had dogged them through the early 1990s and hoped soon to obtain the value of Ω_M that would reveal the universe's future.

Because astronomers had no good observational grounds for assigning the cosmological constant a nonzero value, any curvature of space had to arise from the presence of matter. In that case, the ratio of the average density of matter to the critical density, denoted by Ω_M, must determine the curvature of space and the ultimate fate of the expansion. Observational data suggested a value of 0.2 to 0.4 for Ω_M, not enough to produce the flat cosmos, with Ω_M equal to 1, that the inflationary model of the universe requires. Although these data troubled theorists who believed that the inflationary theory must be correct, most of them assumed that the eventual discovery of still more dark matter would close the gap in omega, establishing the model universe in which $\Omega_M = 1$ and $\Omega_\Lambda = 0$. If $\Omega_M = 1$, however, the universe must have undergone more deceleration as the result of gravity than would have occurred if Ω_M has a value much less than 1. A high-density universe therefore should have an age of only 11 or 12 billion years, which seemed to conflict with the ages of the oldest stars.

These conclusions quaked in 1997 and cracked in 1998, when a new model universe seemed ready to replace the old in the hearts and minds of astronomers and all others who care about cosmology. Although astronomers quite rightly continue to seek possible defects in the recent observations and their interpretation, they have largely shifted their cosmic paradigm. Gone are the happy days with a zero cosmological constant, when the crucial question about a cosmos undergoing a deceleration of its expansion was whether or not the deceleration would eventually stop the expansion and provoke a contraction. In place of those simpler times we find the hurly-burly of a cosmos with a cosmological constant, not decelerating but accelerating its expansion, stripped of all prospect of ever slowing to zero, let alone of recycling its galaxies through a cosmic crunch and possibly into another big bang. Instead, the universe seems destined to an expansion so rapid and so profound that galaxies will eventually stand alone, catacombs of dead stars whose very protons and neutrons will eventually dissolve into a sea of ever-duller radiation.

Let us pull ourselves from the slough of despond in which the implications of the new observations threaten to drown us. Rather, let us lift our spirits by celebrating the astronomical powers of insight that have brought us the latest news about the universe.

SUPERNOVAE AS STANDARD CANDLES FOR
AN IMPROVED HUBBLE DIAGRAM

The discovery of the runaway universe arose from astronomers' improved abilities to observe and to interpret the light from stars that have exploded in galaxies billions of light-years beyond the confines of the Milky Way. For seven decades, astronomers have striven to extend the Hubble diagram upward and to the right, to greater distances and greater redshifts, the result of larger recession velocities. They have done so partly from a basic desire to learn more about the universe but far more because they have known that a sufficiently accurate Hubble diagram, extended to sufficiently large distances and redshifts, will reveal the future of the universe. The revelation would occur because the line on the Hubble diagram showing the relationship between distances and redshifts would establish one model universe as the real one, while stamping its rivals with the label "theoretically possible but practically unrealized."

As we have seen, if a number of objects all have the same luminosity or intrinsic brightness, they will furnish astronomers with a much-needed set of standard candles. Because the apparent brightnesses of these objects decrease in proportion to the square of their distances, astronomers can determine the ratios of distances to the objects simply by measuring their apparent brightnesses. When they plot the Hubble diagram, which shows how the objects' apparent brightnesses decrease as their redshifts increase, the diagram displays a relationship that depends on the values of Ω_M and Ω_Λ. Thus a sufficiently accurate Hubble diagram, based on reliable standard candles, will reveal the prime mysteries of the universe: the average density of matter and the size of the cosmological constant. In this effort, supernovae have carried the palm in the final years of the century.

Supernovae—stars that explode as they reach the final stages of their stellar lifetimes—achieve enormously high luminosities for a month or two, a fact that allows astronomers to study them even when they appear in galaxies many billion light-years from the Milky Way. If all supernovae reached the same maximum luminosity, they would fully satisfy astronomers' requirements for a set of standard candles. Unfortunately for those who dream of such explosive simplicity, supernovae arise from a variety of causes and display a set of different characteristics, including their peak luminosities. This fact has created a modest supernova industry, including a few dozen astronomers who observe stellar explosions and a few dozen more who attempt to calculate how and why these explosions occur. Although we shall spend the next two chapters exploring the trials and successes of these supernova observers and theorists, for now we may skip past most of their problems to emphasize how their efforts have changed cosmology.

Fundamentally, supernovae arise from one of two distinct causes. Either a star's core collapses after losing all ability to produce new kinetic energy by nuclear fusion, or an explosion occurs throughout a white-dwarf star, which consists of unstable matter that physicists call "degenerate." Both types of supernova explosions release fantastic amounts of energy in the form of light and other types of electromagnetic radiation, as well as in streams of neutrinos. By studying the details in the spectra of the light emitted by supernovae, astronomers came to recognize two basic patterns, which they named "Type II" (for the core-collapse events) and "Type I" (for the degenerate-matter explosions). Eventually, they created subclassifications of these types and realized that only the Type Ia supernovae arise when white dwarfs explode their degenerate matter. Although theorists cannot fully explain how these explosions occur, supernova observers have discovered a crucial important cosmological fact about Type Ia supernovae: They all reach nearly the same maximum luminosity. Chapter 9 presents the history of this discovery, including the details that make astronomers believe in its accuracy. For now, we can summarize their efforts with a sensationalistic but largely correct statement: *Type Ia supernovae provide by far the finest, most luminous standard candles that astronomers have ever found to determine the distances to faraway galaxies.*

Happily, when astronomers observe a distant supernova, they can recognize which type of supernova they have detected, despite the fact that the Doppler effect arising from the expansion of the universe has increased all the wavelengths and decreased all the frequencies in the spectrum of the supernova's light. Because the Doppler effect changes all the wavelengths and frequencies in the same proportion, the complex patterns in a supernova's spectrum remain unchanged. After a few years of training on the spectra of bygone supernovae, a supernova expert can recognize these spectral patterns as easily as an automotive expert can identify a 1958

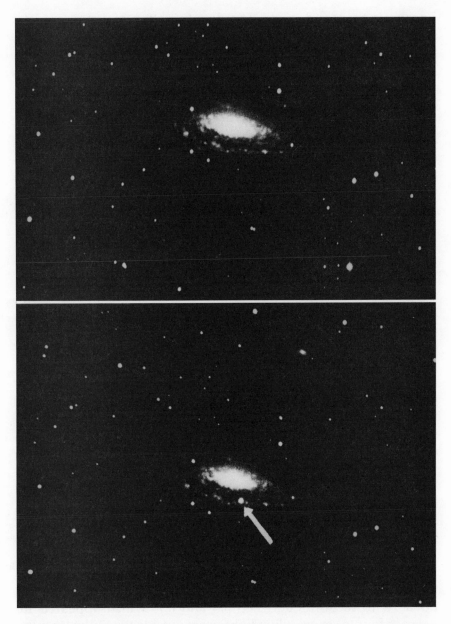

A supernova in a relatively nearby galaxy produces a new, easily visible point of light, marked by the arrow in the lower panel. Finding, identifying, and measuring supernovae in galaxies hundreds of times more distant confronts astronomers with a serious challenge. (Photograph courtesy of the Lick Observatory of the University of California.)

Edsel at a thousand yards. The determination of a supernova's redshift follows directly from measuring the ratio of the wavelengths of the features observed in its spectrum with those produced by the same elements observed in laboratories on Earth.

Before comparing the apparent brightness of a distant supernova with the brightnesses of supernovae observed at much closer range, astronomers must allow for the fact that a substantial redshift in the spectrum will significantly alter the portion of a supernova's output that appears as visible light. Consider, for example, a supernova observed at a redshift of 0.5, the spectral pattern of which shows wavelengths that are 50 percent greater than they would be in the absence of any recession velocity. For this supernova, the Doppler effect has shifted what would be the blue portion of the spectrum if no redshift existed into the red spectral region, and it has moved what would be the red part of the spectrum well into the infrared. Because supernovae do not emit the same amount of light in each broad region of the spectrum, astronomers must adjust their calculations to allow for these redshifts, which they can determine quite independently of the supernova's apparent brightness. Thus, for instance, supernova observers often compare the apparent brightness of a relatively nearby supernova in the blue region of the spectrum with that of a distant supernova observed in the red. To compare a nearby supernova's apparent brightness in the red with the same spectral region for a distant supernova, astronomers must observe the distant object with an infrared camera, often the NICMOS instrument operated on the Hubble Space Telescope until the end of 1998. The emission and absorption of infrared radiation within the Earth's atmosphere prevents ground-based observatories from observing much of the light from a distant supernova, which the cosmic expansion has shifted from red into infrared wavelengths.

Generations of supernova experts, having honed their skills on relatively nearby supernovae, turned their gaze outward during the 1990s, discovering and measuring dozens of Type Ia supernovae with redshifts between 0.3 and 0.7, which appeared in galaxies four to seven billion light-years outside the Milky Way. By devoting years of time, effort, and experience to these supernovae, they learned how to make the minor corrections to each of them that would allow each to serve as a standard candle, capable of providing an accurate distance estimate, as well as a well-determined redshift. With the obstacles overcome and the measurements made, the supernova experts could revel in their new and improved Hubble diagram. This diagram amounts to a road map of the cosmos, not in space, but in time. By learning to read this diagram like an astronomer, the reader may share the joy of discovery and the heartbreak of the cosmological constant.

THE NEW AND IMPROVED HUBBLE DIAGRAM

Like its predecessors, the new Hubble diagram, which we encountered earlier in Figure 4.4, displays the redshifts of standard candles—the Type Ia supernovae—along its horizontal axis (see the top portion of Figure 7.1). Notice, however, that the redshift steps proceed by multiplication rather than by addition, so that the same amount of space along the axis that takes us from a redshift of 0.01 to 0.1 also takes us from 0.1 to 1. A redshift of 1 means that all the wavelengths in the spectrum have doubled and that all the frequencies have fallen by half, in comparison with their values in the absence of any recession velocity. This redshift corresponds to a recession velocity equal to 60 percent of the speed of light. Larger redshifts would more than double the wavelengths and would reduce the frequencies by more than half, so that a redshift of 2, for example, arises if the recession velocity reaches 80 percent of the speed of light. For now, however, even a redshift of 1 lies beyond the observers' accurate reach. The vertical axis in the new Hubble diagram shows the estimated distances to the supernovae, measured in parsecs, each of which equals 3.26 light-years. One megaparsec, or one million parsecs, therefore equals 3.26 million light-years.

So much for axes that form the lineaments of the Hubble diagram. What does the diagram tell us? Each point on the diagram represents a supernova whose redshift and distance have been carefully determined. The lower left-hand portion of the diagram shows both that supernovae provide good standard candles and that Hubble had a pretty good law when he wrote $v = H \times d$: The redshifts, which correspond directly to velocities for relatively nearby supernovae, increase in a straight-line relationship with increasing distance. Because the distances scale with the square roots of the apparent brightnesses, and because the graph plots the redshifts and distances in a multiplicative rather than an additive

FIGURE 7.1   Hubble Diagram Showing Recent Supernova Observations

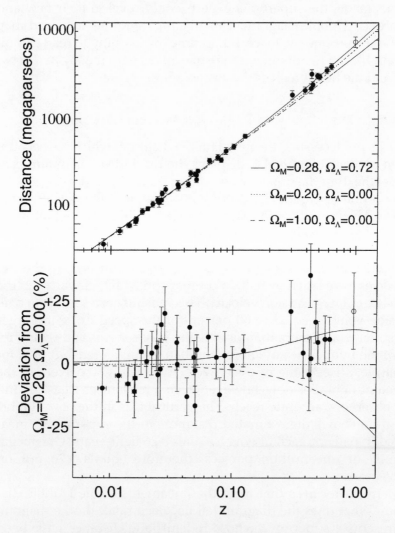

The top portion of this diagram plots observations of supernovae in distant galaxies, with the most distant supernovae at the upper right. The different lines denote the predictions of different model universes, characterized by different values of Ω_M and Ω_Λ. The lower portion of the figure presents the observations as deviations from a universe in which $\Omega_M = 0.2$ and $\Omega_\Lambda = 0$, and expands the vertical axis, so that the observational error bars become strikingly evident. (Diagram courtesy of Dr. Adam Riess and the High-Z Supernova Search Team.)

manner, the straight line through the observed supernovae in the lower left-hand portion of the diagram implies that their recession velocities vary in direct proportion to the distances. This straight-line relationship well describes the supernovae up to redshifts close to 0.1. At that point, we encounter subtle deviations, harbingers of a cosmological revolution.

A CLOSER LOOK: SMALL DEVIATIONS CAN REVEAL A UNIVERSE WITH A COSMOLOGICAL CONSTANT

What are the details that reveal to astronomers the cosmological model that describes the real universe? Drawn on the top portion of Figure 7.1 are three slightly curving lines, each of which represents what astronomers expect from a different model universe. The top solid line denotes a universe in which Ω_M equals 0.28 and Ω_Λ equals 0.72, so that their sum equals 1 and space in the universe is flat. The central dotted line delineates a universe with Ω_M equal to 0.2 and the same value of Ω_M but an Ω_Λ equal to 0; the bottom dashed line corresponds to a flat universe in which Ω_M equals 1 and Ω_Λ equals 0. Until we reach redshifts greater than 0.1, these three lines are almost identical. Only toward the top right-hand portion of the diagram can we see any differences, and they do not allow us easily to distinguish which model holds the real universe in its grasp.

To help portray the differences between the model universes, the experts have drawn an expanded version of the top right-hand portion of the Hubble diagram (see the lower portion of Figure 7.1). In this enlarged version, the model universe with $\Omega_M = 0.2$ and $\Omega_\Lambda = 0$ has been used as a reference, with the effects of increasing distance on apparent brightness removed. In this benchmark model, the brightness ratios of distant supernovae run horizontally, because the vertical axis plots only the difference between the observed values and those expected for the model universe with $\Omega_M = 0.2$ and $\Omega_\Lambda = 0$. This method of presenting the data makes it easier to compare other models, as well as the actual observations, with the benchmark model.

In the expanded version of the most distant portion of the Hubble diagram, the first characteristic that strikes the eye consists of the sizable error bars that extend above and below each point representing a single supernova. In contrast to their redshifts, which can be measured with an accuracy of 1 percent or better, the apparent brightnesses can only be estimated, with possible errors, expressed in the vertical error bars, of plus or minus 15 percent. No single supernova could allow astronomers to discriminate among the three models, because the error associated with estimating its apparent brightness is as large as the divergence among the models. However, after measuring a number of supernovae at large distances, astronomers can apply a statistical analysis that reduces the effective errors in the distance estimates that distinguish each model from the others. This analysis assumes that the errors are random; that is, the chances are equal that a supernova's actual apparent brightness lies slightly above its measured value or slightly below it. For a sizable number of supernovae, the effects of these errors tend to cancel one another, so that the complete data set can provide a much finer means of discriminating among models than a single supernova can. By analyzing the full set of supernovae, including the supernovae with redshifts between 0.3 and 0.7, astronomers reached conclusions that rest on a statistically trustworthy foundation. As will be described in Chapter 10, astronomers now worry not about observational errors, but rather that some systematic difference between relatively nearby and much more distant supernovae might still be misleading them into the shocking conclusion that they presented to the world in 1998.

That conclusion, visible in the right-hand portion of the expanded version of the distant portion of the Hubble diagram (Figure 7.1), is this: *The redshifts and apparent brightnesses of Type Ia supernovae reject the possibility of a flat universe with a zero cosmological constant and suggest a flat universe with an average density of matter much less than the critical density.* The interpretation of the data rests on the fact that the high-redshift supernovae are more distant than they would be in a flat universe with a zero cosmological constant; as a result, their points on the expanded diagram lie well above the dashed, downward-curving line that denotes what we would expect from such a cosmos. In addition, since the observational points also lie above the flat line that describes a universe with a cosmological constant equal to 0 and Ω_M equal to 0.2, they tend to exclude a universe in which all the curvature of space arises from matter, that is, a universe with a cosmological constant equal to 0 and Ω_M equal to about 0.2. This finding does not have as much statistical weight as the first one does, because the points are not as far from the flat line depicting a cosmos with $\Omega_M = 0.2$ and $\Omega_\Lambda = 0$

as they are from one with $\Omega_M = 1.0$ and $\Omega_\Lambda = 0$. Our familiarity with these parameters allows us to examine and to enjoy a new type of diagram that astronomers invented to help them discuss the results of the supernova observations.

'PLOTTING THE UNIVERSE IN A NEW WAY

Supernova experts have keenly shared the average citizen's unwillingness to revise our conception of the universe on the basis of a few observational points, each of them burdened by a sizable error bar that arises from the difficulty of estimating the distances to supernovae billions of light-years away. In order to examine and to understand the results of their observations, the supernova experts have employed a new type of graph, one that displays values of Ω_M along its horizontal axis and values of Ω_Λ in the vertical direction (see Figure 7.2). On this graph, every point represents a particular model universe, characterized by two parameters, the values of Ω_M and Ω_Λ. Thus every point on the graph of Ω_M versus Ω_Λ corresponds to a particular line on the Hubble diagram, which shows the effects on the expansion of the universe produced by matter's gravitational attraction (Ω_M) and the tendency of the cosmological constant (Ω_Λ) to make the universe expand ever more rapidly.

Because the value of Ω_M specifies the amounts by which gravitation has slowed the expansion and the value of Ω_Λ describes how the cosmological constant has accelerated it, the two parameters jointly determine how well the value of $1/H$ provides the age of the universe. In a universe in which Ω_M and Ω_Λ were both equal to zero, that age would exactly equal $1/H$, or about 15 billion years. Even before we examine the observational data, we may admire a bare plot of Ω_M and Ω_Λ, which presents the range of possible ages for the cosmos. On this graph, lines of each particular age appear as roughly straight lines, running from the lower left to the upper right. In the upper left-hand portion of the diagram, the cosmological constant has such a large value, in comparison to the density of matter, that no big bang could ever have occurred, because the cosmological constant would have prevented all of space and matter from ever producing a situation of nearly infinite density. This part of the diagram has been included for completeness; if the actual values of Ω_M and Ω_Λ put the cosmos in this region, cosmologists would have to resign their degrees and find new jobs.

88

FIGURE 7.2   Plot of Ω_M Versus Ω_Λ, Showing Contours of
Constant Age of the Universe

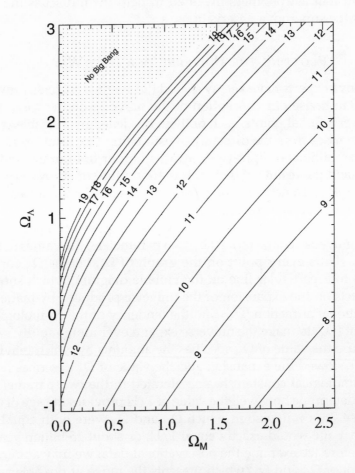

If we make a graph with the vertical axis showing values of Ω_Λ and the horizontal axis showing values of Ω_M, the actual universe can be represented by a single point. Cosmologists have devoted considerable effort to determining the location of that point, a location that remains uncertain. We do know, however, that every point on the diagram corresponds to a particular amount of time since the big bang; these amounts, measured in billions of years, appear as lines on the diagram that run from lower left to upper right. Sufficiently large values of Ω_Λ would imply that no big bang had ever occurred—quite a conundrum for cosmology, should observational results lead us into that region. (Diagram courtesy of Dr. Adam Riess and the High-Z Supernova Search Team.)

THE RESULTS FROM OBSERVATIONS OF
TYPE IA SUPERNOVAE

On the graph of Ω_M versus Ω_Λ (see Figure 7.3), the results from the supernova observations of the mid-1990s appear as elongated ellipses, each of which shows the statistical likelihood that the actual universe lies within a particular ellipse, defined by a statistical level of confidence. The diagram showing Ω_M versus Ω_Λ depicts these ellipses for three levels of statistical confidence, with the largest ellipse almost certain (997 times out 1,000, if we had 1,000 different chances to observe it) to contain the real universe. This diagram statistically crushes the last faint hope of those who have believed in a flat universe with a zero cosmological constant. The point with $\Omega_M = 1$ and $\Omega_\Lambda = 0$ lies far outside the largest ellipse, which contains all but 0.3 percent of the statistical likelihood: Less than 1 chance in 300 exists that the true values of Ω_M and Ω_Λ do not lie somewhere within this ellipse, provided that the observations do not suffer from undetected systematic errors of the sort we shall discuss in Chapter 10.

Figure 7.3 also shows the prediction made by the inflationary model: that Ω_M and Ω_Λ sum to 1. This prediction appears as a straight line that starts at the point on the left-hand vertical axis where Ω_M equals 0 and Ω_Λ equals 1, extends downward and to the right, passing through the point where Ω_M equals 1 and Ω_Λ equals 0. In a universe without a cosmological constant, the inflationary theory predicts that the latter point specifies the true universe. If we allow the possibility of a nonzero cosmological constant, the inflationary model predicts that the actual universe must be represented by a point somewhere along this line, all the points of which denote flat space, neither positively nor negatively curved. For values of Ω_M and Ω_Λ that locate the real universe somewhere above and to the right of this line, space must be positively curved, so that the universe curves back on itself. Below and to the left of the flat-universe line, space must be negatively curved, with the result that travel in a straight line will never bring a voyager back to her starting point.

Why do the error-bounding regions in Figure 7.3 have the shape of elongated ellipses, bounded on the left by the vertical axis, instead of being circles? What property of the supernova observations prevents these regions from being nearly circular, as would

FIGURE 7.3   Plot of Ω_M vs. Ω_Λ, Including Statistical Confidence Boundaries from Observational Results

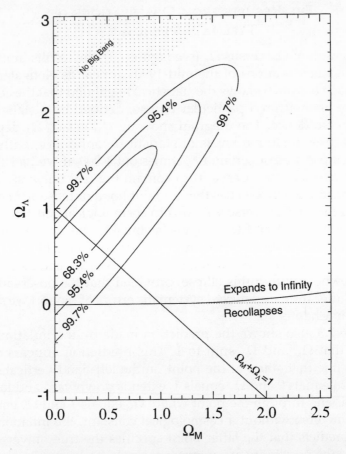

The analysis of the recent supernova observations made by the High-Z Supernova Search Team designates the region of the diagram most likely to include the actual values of Ω_M and Ω_Λ. In the diagram, the contours show the regions that are likely to include the actual universe with probabilities of 68.3, 95.4, and 99.7 percent, provided that no systematic source of error has confused the data analysis. Note that these contours easily allow for the possibility that $\Omega_M + \Omega_\Lambda = 1$ but exclude the possibility that $\Omega_\Lambda = 0$ unless Ω_M has an extremely low value of 0.1 or less, which is highly unlikely because other observations imply that Ω_M equals at least 0.2. These results, obtained by one group of supernova observers, agree with those found by another group, shown in Figure 9.4. (Diagram courtesy of Dr. Adam Riess and the High-Z Supernova Search Team.)

be true if the observations yielded values of Ω_M and Ω_Λ separately? The answer to this seemingly complex question sheds a useful light on astronomers' quest to determine the key cosmic parameters. Observations of distant Type Ia supernovae reveal neither Ω_M nor Ω_Λ alone, but rather their *difference*, $\Omega_M - \Omega_\Lambda$. Why is this so? The energy density provided by the matter in the universe slows the expansion of the cosmos, whereas the energy density implied by the cosmological constant provokes more rapid expansion, pushing different regions apart at an increasing rate as new space comes into existence, complete with additional energy that drives the expansion rate to still-greater values. Although the gravitational effect from matter and the effect of energy from the cosmological constant curve space in the same way, the two effects act in opposite ways on the expansion of the universe. In reaching a complete overview of the effects that slow the cosmic expansion, the two key parameters, Ω_M and Ω_Λ, therefore appear with opposite signs. As a result, observations of the deviation from a straight line in the Hubble diagram can reveal only $\Omega_M - \Omega_\Lambda$ (or, if we prefer, $\Omega_\Lambda - \Omega_M$), but neither one independently of the other.

Analysis of the supernova data shows that the likeliest value for $\Omega_M - \Omega_\Lambda$ equals approximately –0.4. Within the largest of the error-bounding ellipses, the greatest value for $\Omega_M - \Omega_\Lambda$ equals 1 and the smallest equals –0.3. Observations of the motions of galaxies in clusters, and of the distribution of clusters in space, imply that Ω_M must be greater than 0.2, probably lies closer to 0.3, and almost certainly does not rise higher than 0.5. In that case, using the largest error ellipse leads to the conclusion that although Ω_Λ still might be 0, this can occur only if Ω_M has a relatively low value, equal to 0.2 or less.[1] Even this result lies outside the smaller error ellipses, implying that the possibility of a zero cosmological constant, while not yet excluded, must be judged unlikely. Further observations of distant supernovae may soon eliminate this possibility entirely, as has already happened if we believe that the sum of Ω_M and Ω_Λ must be close to 1.

In Figure 7.3, the difference of Ω_M and Ω_Λ has a constant value along any line that runs upward and to the right at a 45-degree angle. These lines are nearly the same as the lines in Figure 7.2 that specify the time since the big bang, for the excellent reason that the

---

[1] The reader will be glad to learn that we are somewhat simplifying the interpretation of the data in this paragraph, without seriously compromising its accuracy.

same effects compete in similar ways to determine both the deviations from a constant rate of expansion and the time that has elapsed since the expansion began. The recent supernova observations therefore provide a helpful discriminant (still not as accurate as astronomers would like) among different values for the age of the universe. They suggest a most likely age close to 14 billion years, and they eliminate ages less than 11 billion years or greater than 18 billion years. This fits well with what astronomers know about the ages of stars and galaxies. The lowest of the possible ages, close to 11 billion years, would provoke some serious head scratching, but most astronomers grow bald rapidly in any case.

In mid-1999, astronomers employing other techniques to measure the distances to galaxies announced that the universe has an age of only 12 or 13 billion years. Instead of analyzing these observations in detail, we may use them to emphasize the salient fact that determining the age of the universe remains a crucial task, to be accomplished at a high level of accuracy by astronomers of the new millennium. For now, we may congratulate ourselves that astronomers have found the time since the big bang (so most of them believe!) within 1 or 2 billion years and may set that time at 14 billion years without doing serious damage to our understanding of reality. The great cosmological news of the late 1990s resides not in finer estimates of the time since the universe began its expansion, but in the apparent discovery that the expansion is accelerating.

COULD THE ACCELERATION ARISE FROM EFFECTS OTHER THAN A COSMOLOGICAL CONSTANT?

Those who learn that the cosmos now appears to be accelerating its expansion oftentimes ask, Why must this acceleration arise from a cosmological constant, an invisible energy that permeates empty space? Couldn't some other aspect of the universe, possibly one less mind-bending than the claim that a constant amount of energy exists in every cubic centimeter of empty space, produce the acceleration?

Broadly speaking and translating from technical language, the answer is no. Einstein softened up the cosmological community, one might say, by introducing the cosmological constant, which provided a tendency toward acceleration that balanced gravity's tendency to produce a cosmic contraction. His analysis of the equations that he derived from his general theory of relativity demonstrated to the satisfaction of his colleagues and their successors that unless we are prepared to reject the tenets of general relativity theory, the only additional term that may appear to produce an acceleration consists of exactly what we have discussed: a cosmological constant of unknown size, but one with the properties we have described—namely, a transparent energy, untappable and untouchable, so far as we can tell, except by its tendency to make the universe expand more rapidly.

Of course, Einstein might have been wrong on this point. Most people are, at least some of the time. However, as generations of physicists and cosmologists grappled with Einstein's equations, attempting to tease out their meaning and to test their validity, some of them devised variations that would revolutionize our understanding as deeply as Einstein ever did—if they prove to be true. To the extent that scientists have devised tests of Einstein's general theory of relativity, however, the champion has prevailed against all challengers. Some of the challenging theories, for example, predict that objects with mass will bend space in slightly different amounts than Einstein's theory predicts. Extremely accurate measurements of the gravitational deflection of radio waves by the sun and moon have verified Einstein and rejected all others, to the limits of experimental accuracy. This does not mean that Einstein must be right, only that he is doing just fine; until a better theory comes along—one that explains more observations with a greater coherence, as scientists judge these things—we should take Einstein's theory as correct, reserving the right to change our minds when and if the evidence swings against it.

THE NEXT BIG ISSUE:
FINDING Ω_M AND Ω_Λ SEPARATELY

Astronomers' observations of supernovae have revised fundamental ideas about the cosmological constant by demonstrating that Ω_M minus Ω_Λ has a value close to –0.4. Most observationally oriented astronomers had expected this difference to lie close to 0.3, because Ω_M had been deduced to be close to this value, and no good reason existed for a nonzero value of Ω_Λ. Theoretical cosmologists expected this value to lie closer to 1.0, because their theories implied that Ω_M plus Ω_Λ should equal 1, and they had no reason to conclude that Ω_Λ should not be 0. Both groups of astronomers thus found themselves astounded, if not totally floored, by what supernovae revealed about the value of $\Omega_M - \Omega_\Lambda$. They have, however, leaped from the carpet, dusted themselves off, and proceeded to investigate the universe.

Cosmologists, observational astronomers, and even average citizens would like most of all to know neither the sum nor the difference of Ω_M and Ω_Λ, but rather the values of each of these terms separately. This would tell us both the total density of matter and the size of the cosmological constant. By themselves, the current supernova observations cannot provide these two numbers individually. Fortunately, however, astronomers have other means to estimate Ω_M and Ω_Λ, either individually or in combination, which we shall examine in Chapters 11 through 14. Before we do so, and before we pay homage to the amazing story behind the discovery that Ω_M minus Ω_Λ equals approximately –0.4, let us turn our attention to the revolutionary implications of living in a universe with a nonzero cosmological constant, as now appears to be likely.

THE SIGNIFICANCE OF
A NONZERO COSMOLOGICAL CONSTANT

The supernova results announced in 1998 imply that the difference $\Omega_M - \Omega_\Lambda$ equals approximately –0.4. Thus, if Ω_M equals about 0.3, as other observations suggest, then Ω_Λ must be close to 0.7. In that case, Ω_Λ is more than double Ω_M, which means that the cosmological constant has twice as large an effect on the expansion of the universe as does all the matter that the mighty cosmos contains!

Quite understandably, in view of the fact that most cosmologists had concluded that the cosmological constant should be zero, this result initially provoked more skepticism than speculation. Only after astronomers had grown reasonably satisfied that the supernova experts had taken every precaution to avoid errors in obtaining their observations and in interpreting them could they accept the notion of a nonzero cosmological constant Λ, a hidden energy in empty space. Their conservative approach, always appropriate in science, received psychological support from the most evident cosmic consequence that flows from a nonzero cosmological constant. If Λ and Ω_Λ are nonzero, the universe must be accelerating. In that case, the expansion will never stop, but in fact will proceed more and more rapidly, eventually driving everything in the cosmos so far apart that even to observe other clusters of galaxies will become technologically impossible. For those who hope that our descendants, hundreds of billions of years in the future, will travel between galaxies to join in a single cosmic civilization, the nonzero cosmological constant represents terrible news. For the rest of us, a nonzero value of Ω_Λ implies that the universe will never cease its expansion, nor undergo a "big crunch," nor recycle itself through another big bang (a highly speculative outcome even if a big crunch were to occur). We must make the best use of our single cycle of cosmic expansion, which appears destined never to end.

THE KERRIGAN PROBLEM

The supernova discoveries also raise a deeply puzzling issue, to which we shall return after assessing the best observational evidence to provide accurate values of Ω_M and Ω_Λ. What counts most in the cosmological arena is not whether Ω_M exceeds Ω_Λ or the converse, but rather the fact that Ω_M and Ω_Λ have roughly equal values. Why is this so? *If Ω_M and Ω_Λ have even approximately equal values, we live at an extraordinary time in the full history of the universe.* In a universe with a nonzero cosmological constant, Ω_Λ has been increasing ever since the big bang, while Ω_M has been decreasing. Soon after the big bang, Ω_M was so much larger than Ω_Λ that a model of the cosmos with Ω_Λ equal to zero described the universe quite accurately. No one alive in those eras (and no one was) could or did assert that Ω_M and Ω_Λ had similar values. In the far-distant future, Ω_Λ will be so much larger than Ω_M that cosmologists will be able essentially to ignore Ω_M. The universe will then behave as if it contained no matter at all, exactly as it does in the model that Willem de Sitter conceived in 1917, and it will become a true runaway, expanding at an exponentially increasing rate.

Only now, speaking in a broad manner, do we find Ω_M and Ω_Λ with approximately equal values. Michael Turner, one of the supernova experts who brought us to this point in astronomical history, likes to refer to the dilemma of explaining why we happen to live in an era when Ω_M and Ω_Λ have similar values as the "Nancy Kerrigan problem." This refers to the famous figure skater who, after an attack instigated by a skating rival, understandably lamented, "Why me? Why now?" In actuality, a simple, though corrupt, explanation exists for the attack on Kerrigan—far simpler than the explanations offered for the approximate equality of Ω_M and Ω_Λ.

Kirshner's Kerrigan problem may seem nonexistent when we learn that if Ω_M and Ω_Λ have approximately equal values now, then they have had roughly equal values for the past few billion years and will continue to have comparable values for 10 or 20 billion years into the future. But this analysis overlooks the fact that a cosmological constant implies that the universe will expand forever at an accelerating rate, so that a time interval of 15 to 25 billion years that includes the present moment amounts to only an infinitesimal fraction of the total history of the universe, most of which is destined to describe an extremely low-density cosmos in which almost nothing happens. In Chapter 14, we shall examine how some cosmologists, venturing into biology and philosophy, attempt to explain the Kerrigan problem with an approach called the "anthropic principle," which roughly states that we would not be here to enjoy our Kerrigan moment if the universe had already come to occupy one of the points that lie far into its infinitely long future.

Before then, however, we should look more deeply into the supernovae that have led to a cosmological revolution. The story of how and why a minority of stars explode at the ends of their lives, while most simply fade into obscurity, has much to tell us about why life exists on Earth, as well as why some forms of life can use these exploding stars to measure the cosmos.

CHAPTER EIGHT

WHY STARS EXPLODE

THE SUPERNOVA PHENOMENON

Throughout history, the stars that explode at the ends of their lives have attracted attention for their sudden, unpredictable appearance amid the basically unchanging patterns of the constellations. Long before anyone understood either the origin of supernovae or their significance in the evolution of the cosmos, astronomers recorded their positions and, in a rough way, even their changing brightnesses. The oldest surviving supernova records come from ancient China, where "guest stars" seemed to portend doom and, according to legend, led to the death of the court astronomers who had failed to predict these apparitions. In the year 1054 C.E., astronomers in China, Europe, and the Middle East followed the rise and fall of the supernova that appeared in the constellation we call Taurus. The ejected residue from this supernova produced the Crab Nebula, the closest and best-studied supernova remnant in the Milky Way.

Within our own galaxy, a supernova appears about once per century, typically rising within a few days' time to an apparent brightness equal to that of the brightest stars in the sky. Since these supernovae typically have distances of tens of thousands of light-years, roughly a thousand times greater than those of the stars that mark the constellations, the familiar rule of apparent brightnesses implies that the supernova luminosities must be millions of times greater than those of stars such as Sirius and

Polaris. Once a supernova passes its peak luminosity, it declines precipitously, fading from unaided eyesight after a few weeks, from detection in an amateur's telescope within a few months, and from visibility with astronomers' giant telescopes after a couple of years.

In 1885, astronomers observed a supernova in the Andromeda nebula, which turned out to be, as Edwin Hubble demonstrated during the 1920s, the closest giant galaxy to our Milky Way. During the first third of the twentieth century, astronomers observed several dozen supernovae in other galaxies, though they have seen none in the Milky Way since two supernovae appeared almost simultaneously (in supernova terms) during the years 1572 and 1604. Astronomers now believe that a well-studied supernova remnant in the constellation Cassiopeia arose from a supernova that appeared three centuries ago within a region so rich in absorbing dust that the supernova passed undetected, or at least unrecorded, by the experts of that era. As the enormous distances to other galaxies became evident, astronomers marveled at the power and luminosity of supernova outbursts, without understanding how any object could release so much energy over a relatively short span of time. Their ignorance seems less surprising when we consider that the fundamental mechanism through which stars produce kinetic energy, the fusion of protons into helium nuclei, remained unknown until Hans Bethe published his theoretical studies of nuclear fusion during the mid-1930s. A crucial insight into how some stars explode, unverified for decades but right on the money, arose in Hubble's immediate neighborhood from the fertile brain of Fritz Zwicky.

An émigré from Europe like Bethe, Zwicky, who had been born in Bulgaria and educated in Switzerland, found a home at Caltech, which gave him employment and an opportunity to collaborate with some of the greatest scientists in the United States. One of these, a highly competent user of the 100-inch telescope on Mount Wilson, was another émigré, Walter Baade. Baade and Zwicky studied the records of exploding stars closely, disentangling two separate phenomena: supernovae and much more modest eruptions, which astronomers called "novae." Since the Latin adjective *nova* means new, the words "novae" and "supernovae" literally mean "new ones" and "super new ones." Novae arise in aging stars that undergo periodic outbursts, so that an individual star may produce novae repeatedly, but supernovae reflect the cataclysmic deaths of stars and can occur no more than once per star. Baade and Zwicky showed that all the observed novae belong to the Milky Way and reach maximum luminosities less than 1/10,000 of those achieved by supernovae. Their conclusion that the two categories originate quite differently has grown ever firmer, so that only the root in the Latin names of the two phenomena reminds us of the fact that both novae and supernovae appear where no star was seen before.

But what could explain supernovae? And could a single explanation suffice to explain all of them? In 1934, Baade and Zwicky made an amazing extrapolation from what physicists had glimpsed about the nature of atomic nuclei; they speculated that "with all reserve we advance the view that a super-nova represents the transition of an ordinary star into a *neutron star,* consisting mainly of neutrons. Such a star may possess a very small radius and an extremely high density." This speculation has proven completely correct for one of the two basic categories of supernovae, and it represents a highly remarkable success in a field where most fine ideas prove to be wrong, the more so as the neutron itself had been detected only two years earlier, in 1932, after being hypothesized to exist on purely theoretical grounds in 1920. In a brilliantly successful leap of imagination, Baade and Zwicky (but almost certainly Zwicky, who had a far greater urge to speculate than Baade did) perceived that the *collapse* of a star's central core could provoke an explosion more violent than anyone had previously conceived that a star could undergo. The oddest fact of all in this successful speculation was that supernovae come in two basic types. All the supernovae that Baade and Zwicky had observed belonged to one of these two types, yet they had deduced the fundamental mechanism about the other type of explosion!

WHEN CORES COLLAPSE: THE ROAD TO
UNDERSTANDING TYPE II SUPERNOVAE

To make the transition from brilliant speculation to an actual understanding of the supernovae that result from a star's central collapse, astronomers had to wait for two separate areas of science to improve significantly. They required both a thorough understanding of all the nuclear-fusion processes that can occur within a star and the ability to perform detailed calculations of how these processes affect its internal structure as the star grows old. World War II and the Cold War that followed gave an enormous impetus to both these efforts, even though the governmental agencies that funded the detailed investigation of nuclear fusion and the development of high-speed computers hardly had the problem of explaining supernovae in mind. During the 1950s and 1960s, a stream of ever-better computers allowed scientists to calculate events and processes far too complex to be investigated mathematically without these fast-acting machines.

An entire class of these complex processes involved hydrodynamics—liquids or gases in rapid motion. One of the hydrodynamical problems most relevant to the Cold War arose in considering an explosion within a mass of gas or liquid, which, in practical terms, dealt with the behavior of nuclear weapons that produced energy by nuclear fusion and released it in Earth's atmosphere. At the great weapons laboratories of the United States near Los Alamos, New Mexico, and Livermore, California, the world's finest computers, in both the human and machine senses, attempted to follow the progression of a furious explosion as it blasted through whatever surrounded it. Some of the scientists involved, eager to investigate natural as well as human-made detonations, turned their attention to using their computer codes to calculate the manner in which an explosion at a star's center would burst its way through the star. This led them naturally to investigate what could cause such an explosion. Nuclear physics then led them back to the original hypothesis of Baade and Zwicky.

Consider what occurs inside a star as it passes from youth to old age. Stars begin their lives made almost entirely of hydrogen and helium, the two lightest and most abundant elements. Brought together by gravitational forces, a star likewise holds itself together by gravity, squeezing itself so tightly that even though it remains gaseous throughout its volume, the gas at its center rises to densities far greater than the density of gold, mercury, or uranium on Earth. The reason that gas at such enormous densities does not solidify rests in the enormous temperatures of the gas. Temperature measures the speeds at which the particles in the gas are moving, and it rises, along with the gas pressure and gas density, as a newborn star's self-gravitational forces squeeze its interior.

NUCLEAR FUSION:
THE KEY TO STELLAR ENERGY PRODUCTION

If this gravitationally induced squeezing told the complete story, a star would collapse immediately. The other half of a star's tale lies in nuclear fusion, which occurs wherever the temperature rises well above 10 million degrees Fahrenheit. At these temperatures, collisions between atoms strip all the electrons from the bare nuclei, and some of the nuclei move with such large velocities that their collisions bring them sufficiently close to one another for them to fuse together, despite the repulsion that arises from the fact that all nuclei carry a positive electric charge. The fusion of nuclei arises from the action of what physicists call "strong" or "nuclear" forces. Strong forces act only over distances comparable to the size of an atomic nucleus, about 10^{-13} centimeter, but within this domain they completely overpower electromagnetic and gravitational forces. Without strong forces, all nuclei would immediately fall apart, leaving our Earth and our bodies a mass of individual protons and neutrons. With them, nuclei can endure indefinitely, and colliding nuclei can fuse together to liberate new forms of energy.

This new energy comes from the mass (more technically, from the energy of mass) contained in the fusing particles. When two protons fuse, for example, they form a nucleus, called a "deuteron," that has a mass about 1 percent less than the combined masses of the protons. A corresponding amount of energy of mass, specified by Einstein's famous formula $E = mc^2$, disappears, to be replaced by an equal amount of kinetic energy, embodied in the speed of the deuteron and of two other particles, a positron and an antineutrino, that emerge from the fusion reaction. Collisions with nearby particles spread this new energy throughout the surrounding gas, so that the kinetic energy produced by nuclear fusion at a star's center diffuses throughout its interior. Some of this energy continuously arrives at the star's surface, heating it to the point that it emits photons of visible light, infrared, ultraviolet, and other forms of electromagnetic radiation. In a stable, middle-aged star, the surface radiates just as much energy per second as the central core produces through nuclear fusion.

Why does fusion occur only at the star's center? Only there do protons move so rapidly that some of their head-on collisions allow strong forces to meld them into a single nucleus. The star's self-gravitation, which squeezes the entire star, has its greatest effect near the center, where it creates a nuclear-fusing core, a giant, natural thermonuclear reactor, which nature has clad in thousands of miles of protective material. From one such nuclear reactor, the Earth receives the light and heat that allow life to flourish.

Poised for Collapse

As nuclear fusion proceeds within a massive star, the star's core eventually comes to resemble a spherical Neapolitan ice cream, with different mixtures of nuclear varieties at disparate distances outward and with the center showing the results from the greatest number of nuclear-fusing reactions. Inside the star, after hydrogen nuclei (protons) fuse into helium, the helium nuclei fuse to produce carbon and oxygen nuclei. Then, as the core grows still hotter and denser, the carbon and oxygen nuclei begin to fuse with helium, soon producing nuclei such as silicon, neon, magnesium, and iron. Each of these fusion reactions turns a progressively smaller fraction of the initial energy of mass into new kinetic energy, so the star, like a misguided drunken motorist, careens ever more rapidly toward catastrophe. The end of the road comes with iron, because the fusion of iron nuclei reverses the energy rule: To fuse iron nuclei absorbs kinetic energy instead of producing it. As the production of iron nuclei comes to an end, the nuclear-fusion processes, hitherto generous providers of kinetic energy, convert themselves into beggars, unable to proceed without a kinetic-energy handout that only gravitational collapse can provide.

As the nuclei in the core of a massive star become mostly iron, the fusion party is over, and the core collapses. In less than a second, a stellar core with more mass than the sun falls inward upon itself in a mighty implosion that produces a neutron star. Squeezed by the infall to a radius slightly smaller than its long-term size, the new-formed neutron star bounces, and this bounce triggers a shock wave that roars outward, reversing the infall of the star's outer parts and blasting them into space at speeds of thousands of miles per second. A small fraction of the stellar material reaches velocities equal to a sizable fraction of the speed of light. The mighty blast initiated by the core's collapse also generates impressive amounts of visible light and other radiation, which soon peaks and then fades away as the ejected material spreads outward. Though the rush of light draws our attention, the supernova's crucial contribution to the cosmos lies in its mulching of interstellar matter with the debris from its explosion. This supernova detritus includes not only nuclei lighter than iron, made before the explosion in relatively large amounts, but also heavier nuclei, fused in small quantities by the blast of the explosion itself.

THE COSMIC LOAM THAT MADE US

When we take inventory of our planet and its teeming forms of life, we find that supernova-made nuclei play a host of crucial roles. Carbon, nitrogen, oxygen, phosphorus, and other light nuclei form the bulk of organic matter. Some of these nuclei may have been expelled from red-giant stars, but most arose in supernovae that exploded long before the sun and its planets formed, 4.5 billion years ago. Nuclei such as aluminum, silicon, magnesium, titanium, and iron almost certainly come primarily from supernovae, and this statement holds absolutely true for the still-heavier elements we prize so dearly, such as silver, gold, mercury, tungsten, and uranium, which supernova explosions long ago flung into a receptive cosmos, a tiny portion of which became our Earth.

THE CHARACTERISTIC SPREAD AMONG TYPE II SUPERNOVAE

The tale of core collapse in massive stars tells the story of Type II supernovae. Because the masses of newborn stars range all the way from one-tenth of the sun's mass up to 50 to 100 solar masses, the progenitors of Type II supernovae exhibit a wide spread in their masses: Any star born with more than eight to ten solar masses represents a candidate for a Type II collapse and explosion. Unsurprisingly, therefore, Type II supernovae exhibit a significant spread in their peak luminosities and energy releases, ranging by at least a factor of ten from greatest to least. This spread eliminates Type II supernovae as standard candles, at least for the straightforward method of comparing peak brightnesses; we might as well attempt to deduce the distance to ships at sea from their apparent angular sizes, without knowing whether we were observing battleships or cabin cruisers. Fortunately, nature has created another class of exploding stars, which reach still-greater luminosities that render them visible at even greater distances, and has also, as we have seen, made them excellent standard candles. What a fine fabricator nature proves to be! In the case of Type Ia supernovae, she weaves so well that we still lack a good explanation of how these stellar explosions occur. Thus, just as Type Ia supernovae have become the darling of observational cosmology, they remain a thorn in theorists' consciences, provoking them to the greater efforts that will yield deeper insights.

THE QUIET ROAD TO DEATH FOR LESS MASSIVE STARS

Like the most massive stars, the sun and other stars with lesser masses will someday exhaust their supplies of protons by fusing them to make more complex nuclei. In about five billion years, when the sun's core begins to run out of protons, gravity will squeeze the sun's core more tightly. This contraction will heat the core and make its remaining protons fuse ever more rapidly, increasing the rate of energy at which the core releases energy. The additional flow of energy outward will expand the sun's outer layers, turning the sun into a "red giant," the cool, distended atmosphere of which will conceal its contracting, nuclear-fusing core. Eventually, the core will grow so hot that helium nuclei themselves will fuse together, releasing more energy—but only about one-tenth as much as the proton fusion that previously made the sun shine. Without a free energy lunch, the sun, like all other stars as they exhaust their original supplies of protons, must either develop new avenues to release kinetic energy by fusion, or find new means of support, or collapse under its own weight.

For the sun and most of its sister stars, physics takes the middle way: The sun will neither perform nuclear fusion after fusing its helium nuclei nor collapse because it has no means of holding itself up. Instead, most stars become degenerate white dwarfs, supported against further contraction or collapse by the effects of the "exclusion principle." Here we meet fancy names for behavior never visible to us on Earth. Degenerate matter, by definition, consists of matter with bulk behavior strongly affected by the exclusion principle. The exclusion principle refers to the counterintuitive fact that certain types of elementary particles—most notably, protons and electrons—refuse to occupy almost the same position with almost the same velocities. The size of the "almost" in this description corresponds to the "quantum" in quantum mechanics and plays a direct role in the ways that electrons orbit an atomic nucleus. In a carbon atom, for instance, the six electrons cannot all move in the smallest allowed orbit, because the exclusion principle allows only two; the other four electrons must occupy the second-smallest orbit or even larger ones. Since we all consist of atoms, the exclusion principle governs our entire lives, even though we cannot observe its effects at the level of sizes that our eyes perceive.

In the core of an aged star, however, we could observe the large-scale effects of the exclusion principle. Once matter packs itself together at densities that approach one million times the density of water, the electrons in the matter will resist further compression, not through conventional notions of pressure, but because the exclusion principle says, No more! In a mixture of electrons plus atomic nuclei, which describes the core of a star that has fused its protons into helium and then its helium into carbon and oxygen nuclei, this suffices to support the entire core. Although the carbon and oxygen nuclei do not feel the exclusion principle directly (one of the principle's many mysteries resides in the fact that it affects only certain types of particles), they do feel the attractive electromagnetic forces between their positively charged selves and the negatively charged electrons. As a result, when the electrons refuse further contraction, so, too, at one remove, do the carbon nuclei. The entire core then sits quietly, no longer capable of nuclear fusion and utterly resistant to compression. As the star's outer layers evaporate into space, the core stands revealed as a slowly cooling white dwarf—white because still hot, dwarf because its size roughly equals the size of the Earth, though the white dwarf contains a starlike mass, several hundred thousand times greater than Earth's.

Squeezed to a density approaching one million times the density of water, white-dwarf material resembles nothing on Earth, but it appears in billions of white dwarfs sprinkled through the Milky Way and many more times over in the galaxies beyond. White dwarfs produce no new energy, for nuclear fusion has ceased forever within them, and they shine because they slowly radiate away the energy bequeathed to them from prior years when fusion occurred. Even the closest white dwarfs to the solar system, only a few light-years away, radiate too little energy to be visible without a good-sized telescope. These dim objects closely mimic how the sun will appear in about six billion years. The great majority of stars, possessing masses comparable to the sun's, will end their lives as white dwarfs, slowly fading into obscurity, eventually becoming "black dwarfs" that form a modest component of the baryonic dark matter.

THE LIMITING MASS OF A WHITE DWARF

Most of the stellar cores that become white dwarfs die quietly thereafter, resisting self-gravitation by turning themselves into degenerate matter. This fate lies in store for the vast majority of stars, all of which have masses less than a few times the mass of the sun. The minority of stars that possess large masses, however, cannot achieve the slow, degenerate death that envelops all others. During the 1930s and 1940s, the brilliant Indian-born astrophysicist Subrahmanyan Chandrasekhar demonstrated that no white dwarf can exist if its mass exceeds 1.44 times the mass of the sun, now called the "Chandrasekhar limit." For reasons that may flow past us as easily as degenerate matter resists further contraction, nature has given the exclusion principle only limited powers: If a mass greater than the Chandrasekhar limit exhausts other means of supporting itself against gravity, it will collapse under its own weight before the exclusion principle can organize proper resistance. The collapse will produce a neutron star, a much smaller object, only as large as a city, in which matter packs together at such enormous densities that a teaspoonful of it brought to Earth would weigh as much as a battleship! Like a white dwarf, a neutron star relies on the exclusion principle to resist collapse, but in this case, the principle acts on neutrons, not electrons, with the result that its effects intervene only at much higher densities.

TYPE I SUPERNOVAE

By the end of the 1980s, astronomers had achieved a fairly complete understanding of Type II supernovae, and they had learned how to estimate their distances, though not with complete success. They were only beginning to understand the other type of supernova, more intrinsically luminous and thus visible at even greater distances, the Type I. Observationally speaking, the most noticeable difference between Type I and Type II supernovae resides in the fact that Type I's show none of the features in their spectra that hydrogen atoms produce. In other words, Type I supernovae apparently lack detectable amounts of hydrogen, a remarkable fact when we consider that hydrogen is by far the most abundant element in the universe. Even aged stars, the cores of which have long since fused all protons (hydrogen nuclei) into heavier elements, have significant amounts of hydrogen in their outer layers, which have never engaged in nuclear fusion.

The total lack of detectable hydrogen in Type I supernovae thus whispered to astronomers that they had found a special class of objects. Eventually, they divided this category into Types Ia, Ib, and Ic, and they realized that a vast difference separates the Type Ia's from the other two. According to our current understanding of how stars explode, supernovae of Type Ib and Type Ic represent variants on the basic core-collapse scenario for massive stars. These supernovae lack hydrogen (the Type Ic supernovae lack helium as well) because they have puffed all of their outer layers into space during their red-giant phases of evolution, leaving behind only the lower layers, in which all the hydrogen (and, in the case of Type Ic's, all the helium also) has been fused into heavier elements. Like their Type II cousins, the Type Ib's and Type Ic's owe their explosions to the moment when their cores collapse, giving up the ghost upon fusing most of the nuclei at their centers into iron.

TYPE IA'S:
THE TRULY DIFFERENT SUPERNOVAE

The preceding analysis leaves the Type Ia's as the luminous exception to the rule that supernovae result from the collapse of the cores of massive stars. Type Ia supernovae do not arise in massive stars; instead, they occur when an explosion rips through the degenerate matter in a white dwarf. This matter is inherently unstable and ripe for explosion. And why is degenerate matter ripe for explosion? In ordinary, nondegenerate matter, such as the nitrogen-oxygen gas that forms the Earth's atmosphere, an increase in temperature produces an immediate expansion of the gas. Warm air rises because the additional heat in the warm gas expands it to a lower density, causing the gas to float higher in the sea of slightly cooler and denser fluid. The direct coupling between the temperature and the density of the gas occurs because the gas molecules can move and interact freely. In contrast, degenerate matter does not respond quickly to a change in temperature, because the particles affected by the exclusion principle have, in effect, ceased to pay attention to the local temperature. In degenerate matter, a rise in temperature will pass unnoticed, since it produces nothing like the change in density that occurs in normal matter.

As an explosion begins in nondegenerate matter, it tends to expand nearby material. For bomb makers, this natural result of a small explosion has always been a problem that threatens to disrupt the rest of the bomb before it can explode. Generations of brilliant scientists and engineers have devised ways to produce a detonation of the entire bomb at a single moment, lest the expansion caused by the first bit of explosion produce a "fizzle." This problem does not arise in a bomb made of degenerate matter, because the material reacts only in the most sluggish manner to anything that happens nearby. Hence an explosive process in degenerate matter has a good chance to roar through the matter before the material "learns" that its surroundings have already exploded.

The carbon nuclei in a white dwarf are prime candidates for nuclear fusion. In massive stars, as we have seen, carbon nuclei fuse to produce heavier elements as an integral part of the fusion chain that leads all the way from hydrogen to iron nuclei. White dwarfs put their feet down on carbon because they have grown so dense that the exclusion principle effectively locks the electrons in place. The electrons, in turn, hold their nuclei by electromagnetic forces, preventing the violent collisions that lead to fusion. However, if a blast wave roars through the white dwarf, the carbon nuclei can fuse to produce heavier nuclei, releasing kinetic energy as they do so. Each fusion releases additional energy, which heats neighboring regions to induce more fusion. Because degenerate matter does not expand quickly in reaction to the release of heat in neighboring regions, a wave of nuclear fusion can spread through the entire white dwarf before it expands. A suitable bomb, which could never explode a normal star, can spark nuclear fusion throughout an entire white dwarf.

SUPERNOVAE IN BINARY SYSTEMS

What could be the bomb that sets off the fusion of carbon in a white dwarf, as a uranium (fission) weapon sets off the fusion in a hydrogen bomb? Astronomers strongly suspect, though they cannot yet prove, that some white dwarfs produce Type Ia supernova explosions because a nearby companion star, passing through the red-giant phase in which its outer layers expand enormously, rains hydrogen-rich material onto the white dwarf's surface. Binary star systems appear in great numbers throughout the Milky Way, and no doubt in other galaxies as well. Half of all the stars that shine may belong to such a system, rather than spending their lives alone, as our sun does. These systems offer the chance for one star to feed the other with the material that will produce a supernova.

Imagine a binary star system whose component stars begin their nuclear-fusing lifetimes with different amounts of mass. The more massive star will fuse its protons more rapidly, maintaining a larger luminosity and reaching its red-giant stage before its sister does. If this star does not have such a large mass that it undergoes a core collapse, it will evolve into a white dwarf many years before its companion becomes a red giant. Then, when the lower-mass star does puff its outer layers into space, the gravitational force from the white dwarf may capture much of this material, still relatively rich in hydrogen nuclei (protons) because the fusion processes in the star did not reach matter so far from the center.

For a time, the gift of material accumulates on the white dwarf, a Trojan horse that will explode the white dwarf from the outside without having to penetrate the white dwarf's degenerate defenses. As more material builds up on the surface, it grows ever denser and hotter. Eventually, the temperature rises well above 10 million degrees Fahrenheit, initiating a blast of proton fusion. The energy from this fusion spreads among the carbon nuclei, overcoming the rigidity imposed by the electromagnetic forces from the electrons and giving the carbon nuclei such large velocities that they fuse together upon collision. In one mighty outburst, the fusion of carbon nuclei throughout the white dwarf unleashes a wave of new kinetic energy, created from the decrease in energy of mass of the fusing nuclei.

The process continues, on a timescale measured in seconds, until nuclear fusion has produced mostly nickel nuclei. These fusion products, and everything else in the supernova, expand into space at high velocity. Unlike a core-collapse supernova, which produces a neutron star (and in the most massive stars, may beget a black hole), the fusion of carbon nuclei in a white dwarf blows the entire object to bits. The light from this type of supernova arises primarily when nickel nuclei decay, producing cobalt and eventually iron nuclei, antielectrons (positrons), and gamma radiation. The gamma radiation fights its way through the explosion's outer layers, changing in part into visible light as it does so. The outer layers, which hinder the rush of energy from below, delay the supernova in achieving its peak luminosity until several days after the explosion has begun.

Can this model for Type Ia's also explain why all Type Ia's reach approximately the same maximum luminosity? To a large extent, it can. We have seen that a white dwarf can exist only if its mass does not exceed the Chandrasekhar limit, just above 1.4 times the mass of the sun. If material from a companion star falls onto a star with a lower mass, it may be fused into carbon nuclei that effectively become part of the white dwarf's degenerate structure, without inducing any explosion. But once a white dwarf's mass has risen to the Chandrasekhar limit, no such accommodation can occur. Instead, matter from a companion star must either cause the white dwarf to collapse, possibly forming a neutron star, or (in what theorists consider the much more likely outcome) set off the fusion that has been described, so that the white dwarf erupts as a Type Ia supernova. In that case, we may have a good explanation of why Ia's reach the same peak energy output: They all come from white dwarfs with masses at the Chandrasekhar limit, hence from the same amount of material ripe for fusion, with the same composition from object to object and the same trigger, the buildup of material that leads to a detonating wave of nuclear fusion that destroys the white dwarf.

This all sounds so reasonable that one may wonder why the finest minds of astrophysics have not yet generated detailed computer models of Type Ia supernovae. They have not been able to do so, in large part because the calculations have proven fiendishly difficult, involving (among other matters) the calculation of how streams of radiation interact with material in violent motion. Experts in these calculations feel some shame—or at any rate, disappointment—from the fact that they cannot tell the observers exactly what to expect when a degenerate white dwarf undergoes sudden, disruptive nuclear fusion. Rather, they must allow observers to tell them what happens in these explosions, at least with respect to a supernova's outer regions, from which we detect its light. Future generations of computers and their gurus will no doubt meet the challenge of creating models that match reality, granting the rest of us confidence in the notion that we understand what makes Type Ia supernovae serve so well as the standard candles of which astronomers dream. Until then, we would do well to concentrate on the observational results from the Type Ia supernovae. These have sufficient importance to merit a sizable chapter, in which we shall meet some of the heroes of the cosmological revolution, the astronomers who have apparently found that the universe seethes with a cosmological constant that will accelerate its expansion forever.

CHAPTER NINE

THE RACE
TO FIND THE FUTURE OF
THE UNIVERSE

THE STUNNING COSMOLOGICAL DEVELOPMENTS of the late 1990s arose in large part from one of the most productive rivalries in the history of cosmology: a race between two teams of astronomers who sought to find the future of the universe by determining the crucial cosmological parameters Ω_M and Ω_Λ. In a Hollywood movie, one of these teams would have excellent funding, well-established scientists, and a tendency to support the conceptual status quo, while the other, destined to early despair but eventual triumph, would consist of misfits who, eschewing traditional methods of research, perceive a truth unacceptable to those whose mind-sets cannot conceive new explanations of reality.

In the world of science, however, the rivalry between two groups of scientists typically follows a different path, one that demonstrates that competition to achieve new insights plays a key role in advancing knowledge. Scientists frequently criticize and mistrust one another, sometimes on a personal level but far more often because they doubt that another scientist has made observations as accurately as claimed or has reached the proper conclusions from what has been observed. Hence scientists repeatedly analyze, assess, and attack any new result announced by their fellows with a vigor that increases in proportion to the significance of its implications. They do so because they know that if they can disprove another's claims, their reputations will rise. This fact embodies the organized skepticism institutionalized in the world of science, the notion that only by surviving harsh criticism from experts can a new result gain respect.

The sleek beauty of the scientific approach to understanding lies in bringing into mutual, institutionally mandated opposition the desires of individual scientists to see their own interpretations triumph. In the short or medium term, much depends on the power and authority of these scientists or on the psychological appeal of a particular interpretation. In the long run, though, precisely because science rewards those who successfully overturn conventional wisdom with the fame that scientists eagerly seek, scientific skepticism brings out the truth—truth as defined by the ability to provide a better explanation of what scientists report in their observations.

The competition between the two teams of supernova experts provides a textbook example of this system at work—so good an example, in fact, that the reader should be skeptical of believing that all scientific discovery proceeds so well. At the price of some ego damage, some moments of harsh talk in private, and some lingering feelings that the other guys stole too much of the glory, the rivalry between the two groups of astronomers who observe high-redshift supernovae brought remarkable new knowledge of the cosmos to the astronomical community and the world. Their success entitles them to our attention as we follow their trajectories toward their hard-won perceptions of the cosmos, which will be especially noteworthy should they prove correct.

THE BERKELEY SUPERNOVA GROUP

The first group of astronomers to become deeply engaged in observations of supernovae with large redshifts has its center in the office of Saul Perlmutter at the Lawrence Berkeley Laboratory (LBL) in California. The LBL, once known as the Lawrence Radiation Laboratory, sprawls on the hillside directly to the east of the campus of the University of California at Berkeley. For generations, the LBL has been a world-renowned center of physics research, where physicists, among their many other activities, have used particle accelerators to accelerate nuclei to energies of billions of electron volts, inducing nuclear-fusion reactions to produce a series of heavy elements never seen in nature because they quickly decay into other nuclei. With the creation of ultraheavy, fast-decaying elements such as berkelium and californium, the discoverers of these elements at the "Berkeley Rad Lab" inscribed the name of Berkeley on the periodic table of the elements, and thus on the map of science.

As particle accelerators grew ever larger, more complex, and more expensive, the crucial experiments in high-energy particle physics came to occur not in Berkeley, but in newer, more advanced facilities such as the Stanford Linear Accelerator Center, the Fermi National Accelerator Laboratory near Chicago, and, most important of all, the Center for European Nuclear Research near Geneva, Switzerland. The Lawrence Berkeley Laboratory has become a leading center for physics research other than particle-collision experiments, including environmental, medical, and engineering physics. In addition, since the 1970s, the LBL has maintained a significant presence in astrophysics, in large part because the famous physicist Luis Alvarez, who collaborated with Ernest Lawrence to build particle accelerators during the 1930s and 1940s, became passionately interested in a variety of astronomical problems.

Two decades ago, Alvarez, who died in 1988, collaborated with a team of geologists and geochemists at the LBL and the University of California at Berkeley to explain "mass extinctions" on Earth, such as the famous extinction of the dinosaurs 65 million years ago. In the early 1980s, these scientists, who included Alvarez's son, the geologist Walter Alvarez, and Richard Muller, a young colleague at the LBL, proposed that mass extinctions occur when a comet or asteroid five to ten miles in diameter strikes the Earth, raising a cloud of dust that darkens the skies for months on end. After considerable initial skepticism of exactly the sort mentioned earlier in this chapter, this impact theory of mass extinctions has gained widespread acceptance, at least for many of the extinction events. The dinosaur extinction now rests on the firmest evidence, because geologists have found a large crater at the edge of the Yucatán Peninsula that was made by an impact at the same time that the dinosaurs died. Some alternate theories for mass extinction remain viable, including the suggestion that volcanic eruptions rather than cosmic collisions caused massive climate changes that disrupted the normal possibilities for life on Earth.

Muller, one of Alvarez's closest colleagues during the 1980s, refined the impact hypothesis in his attempt to explain mass extinctions: He proposed that the sun has a dim companion star, a faint
red dwarf that Muller named "Nemesis." In Muller's model,
Nemesis moves around the much more massive sun in a highly
elongated orbit. At its point of closest approach, still hundreds of
times more distant than the sun's planets, Nemesis's gravitational
force affects the orbits of comets, sending some of them much
closer to the sun to produce a "comet shower." This implies that
comet showers, and the mass extinctions produced when one or
more comets strike the Earth, should recur on a periodic basis,
about once every 26 million years.

In addition to investigating whether this periodicity could be established from the fossil record of mass extinctions and the geolog-
ical record of cratering on Earth, Muller naturally wondered
whether he could find Nemesis itself, the closest of all stars (by hypothesis) to the sun but nevertheless a dim object because of its
low intrinsic luminosity. A search for Nemesis would involve surveying the sky in a quest to find a faint object whose shifts in position would reveal its proximity to the solar system: During the
course of a year, as the Earth moved to different positions around
the sun, the parallax effect would give Nemesis the largest displacements of any star (see Figure 4.1). To deal with the immense
task of sifting through all the stars that a telescope would reveal,
Muller and his colleagues developed new computer software that
could guide a telescope robotically and analyze its observations
without human intervention, simply calling attention to those
stars whose changes in apparent position marked them as different from the ordinary. Installed on a medium-sized telescope at the
campus's Leuschner Observatory in the gentle hills to the north-
east of Berkeley, the robotic search program made little progress in
searching for Nemesis (and certainly never found it) before being
converted to a much deeper purpose, the search for supernovae.

Muller saw that his robotic search program could overcome the
great obstacle to finding exploding stars in other galaxies: the
boredom of examining galaxy after galaxy, week after week, to see
whether a new bright object had appeared in any of them. With
improvements in the detectors that received the telescope's light
and in the software to analyze what they saw, the telescope could
perform a prescheduled series of observations on each night, stor-

ing galaxy images in a computer's memory and then subtracting these images from those of the same galaxies taken a few weeks or months later. As a result of this subtraction, any new object in the second exposure would appear as a spot of light on an essentially black background, which the computer could spot far more easily than it could detect an additional point of light within the mass of stars that form a galaxy. By the mid-1980s, Muller's group at the LBL, working in collaboration with astronomers on the Berkeley faculty, had begun to achieve success in their automated super- nova search. At this point, Saul Perlmutter joined the team.

Perlmutter, a slim, wiry, balding man ready to turn forty with the new millennium, resembles Woody Allen with a Ph.D. in as- tronomy, though Perlmutter could probably best Allen in a rapid- fire speech contest. Born in Philadelphia to academic parents who sent their son to Quaker schools for their social consciousness and orientation toward learning, Perlmutter was a Harvard under- graduate who wanted to study both physics and philosophy but dropped the latter in the interest of having a reasonable time in college. Inclined toward the experimental side of physics and ea- ger to address issues of what makes the world work, he began graduate studies at Berkeley with a bent toward particle physics. Before long, he concluded that he wanted to become involved in projects smaller than those typical of experimental particle physics, which are notorious for involving dozens, if not hun- dreds, of scientists and requiring years to complete. Like many physicists, Perlmutter had perceived that astrophysics continually probes fundamental issues, often involving high-energy particles. After encountering Rich Muller and his group of scientists on the hill at the LBL, he was struck by their scientific imagination and flexibility. Muller, who was impressed by Perlmutter's energy, su- pervised his doctoral thesis, which dealt with the techniques in- volved in the robotic telescope search for the hypothetical star Nemesis. Inspired by the possibilities of automated searches, Perl- mutter was happy to be hired as a postdoctoral fellow in the Muller group, and he pressed forward with the search for super- novae—one more physicist who had made the transition to astro- physical research.

Between 1986 and 1989, the LBL's automated supernova search found nearly two dozen supernovae. All these exploding stars lay within relatively nearby galaxies, no more than a few hundred million light-years from the Milky Way. Much more distant super-novae would inevitably be much fainter, making detection much more difficult, if not impossible, with the medium-sized telescope at the local Leuschner Observatory—a telescope that, on the other hand, could be completely devoted to the supernova search. In the late 1980s, Perlmutter and Carl Pennypacker, another member of the LBL supernova group, discussed the fact that supernovae of Type Ia (SN Ia's) had been identified as particularly luminous explosions, detectable at extremely large distances beyond the Milky Way. If SN Ia's could provide standard candles, as some astronomers had suggested, then observations of these supernovae at truly large distances could establish the value of the Hubble constant and, eventually, any deceleration of the universal expansion.

To find even a few of these supernovae per year, the astronomers would have to monitor many thousand distant galaxies, since SN Ia's appear once every few centuries in a large galaxy. Perlmutter and Pennypacker had become experts in the use of charge coupled device (CCD) detectors, which record photons on silicon chips. These chips consist of individual pixels (picture elements), which we may imagine as separate buckets, each of which catches the light from a particular direction. The buckets describe their contents electronically to a computer that builds a digitized image for further analysis. Astronomical breakthroughs were occurring as the CCD industry developed chips containing a sufficient number of pixels (light buckets) to produce a relatively wide-angle image. By 1988, with chips of 1,000-by-1,000 pixels becoming the standard, Perlmutter and Pennypacker could conceive of an automated telescope with a detector that could view several hundred distant galaxies simultaneously. A limited number of such observational areas could provide enough galaxies to make feasible a search for supernovae sufficiently distant to help reveal the key cosmological parameters.

The Berkeley group of supernova observers initially considered naming itself the "Omega Project" after its quest for the value of Ω, which determines the fate of the universe. (In those days omega meant only Ω_M to nearly everyone, since few seriously believed in a nonzero cosmological constant.) A more descriptive name, the "Supernova Cosmology Project," superseded the original notion, but the group's motivation remained clear: the quest to find accurate values of the Hubble constant and of the average density of matter. To do so, the Supernova Cosmology Project would have to find and analyze supernovae at distances of many billion light-years, with redshifts of 0.3 or greater, so that the light from the supernovae would arrive with information about epochs when the universe was significantly younger than it is now.

Within a year, the Berkeley group was hard at work developing the software that would allow its scientists to search hundreds of galaxy images simultaneously, and it had persuaded Harvey Richardson, a designer of optical instruments at Canada's Dominion Astronomical Observatory, to create a novel design that used a mirror rather than refractive glass to focus light from the telescope onto the CCD detector. By using reflection instead of refraction, the optical system could avoid distortions in color that refraction invariably produces. At just about this time, astronomers detected the first high-redshift supernova of Type Ia—not through the Berkeley group's work, but as the fruit of the efforts of a small group of European and Australian astronomers.

THE DANES BLAZE THE PATH TO DISTANT SUPERNOVAE

During 1987 and 1988, a team of astronomers from Denmark, Australia, and the United Kingdom, led by the Danish astronomers Hans-Ulrik Nørgaard-Nielsen, Leif Hansen, and Henning Jørgensen, used a relatively small telescope in the Canary Islands, equipped with a new CCD detector, to monitor about sixty distant clusters of galaxies, each containing many dozen large galaxies. In August 1988, a supernova appeared in one of these galaxies with a spectrum, analyzed with the largest telescope in Australia, that revealed the characteristic features of a Type Ia supernova, seen at a redshift of 0.31. As the Danish-led group correctly noted, if a dozen or more such supernovae could be found at similar redshifts, and if Type Ia supernovae could serve as good standard candles, then astronomers could derive the value of omega from observations of high-redshift SN Ia's. To achieve this result, the group calculated that it would have to follow more than a thousand galaxies for several years. Had the Danish astronomers continued their efforts and accomplished this goal, the race to find the key cosmic parameters would not have become an essentially

American affair. However, the Danes saw that at their rate of finding supernovae, accomplishing this goal would require many decades, since they had found only a single high-redshift SN Ia in two years of observations. Late in 1988, the Danish astronomers submitted a proposal to use the Hubble Space Telescope, then nearing its launch date, to find distant supernovae. When the Hubble Space Telescope's Time Allocation Committee rejected this proposal, the Danes concluded that they had neither the wherewithal nor the prospects of sufficient telescope time to justify spending several decades on the quest for high-redshift supernovae. So the Danish astronomers, having made the initial discovery that showed the feasibility of the approach, turned their attention to other astronomical pursuits.

THE QUEST FOR FUNDING IN
THE UNITED STATES

In the United States, securing funding was a serious problem for the Berkeley supernova group, which had yet to find a single distant supernova. Fortunately for its efforts, the National Science Foundation (NSF) had developed an interest in creating research centers throughout the United States. At the University of California at Berkeley, physicists and astronomers, including many of the astrophysicists working at the LBL, proposed a Center for Particle Astrophysics (CfPA), which came into existence as the 1980s ended. For a time, the CfPA's budget provided half of the funds to continue the automated supernova search, and the LBL, the remainder. To secure continuing funding, however, the supernova search would have to justify itself to the physics section of the NSF, as well as to the LBL and the CfPA. All of these institutions, in time-honored fashion, appointed committees to review the feasibility and progress made by the Supernova Cosmology Project. Although some of the reviewers doubted that the search technique would actually succeed, the overall reports were favorable, and the head of the physics section of the NSF became a strong supporter of the project. This support was sorely tested during the first years of the 1990s, as the Berkeley group attempted to prove its worth without real success to report.

During 1990, while the Berkeley supernova group developed the software for its supernova search, Warrick Couch, an astronomer at the Anglo-Australian Observatory near Coonabarabran, Australia, who had worked with the Danish-led group, built the electronics to record the wide-angle images of galaxy clusters on CCD detectors. The astronomers were then ready to record data and to analyze it for possible supernovae. They lacked only one crucial opportunity: observing time. Every large telescope receives far more requests for use than can possibly be granted, so a "time allocation committee" must decide, with all the strife this can provoke, which projects receive observing time at the expense of others. In this competition, observing programs with prior success carry clout with the committee, while innovative, unproven approaches must surmount a catch-22: how to obtain the observing time needed to demonstrate the technique's feasibility without any accomplishments to report.

Because the Berkeley astronomers' techniques were largely unproven, the Anglo-Australian Observatory granted them only twelve observing nights over a period of one and one-half years, nine of which proved to be cloudy, as often happens in Australia. Although Perlmutter and his colleagues demonstrated that their software functioned properly, five of the six potential supernovae the group found turned out to be active galactic nuclei or quasars. In 1991, the Berkeley supernova group seemed likely to suffer the same fate as the Danes, capable of finding high-redshift supernovae but incapable of exciting enough interest to be supported sufficiently to find the dozen or more supernovae that would yield cosmologically significant results.

Seeking better observing conditions for the supernova search, Perlmutter established a collaboration with British astronomers who were using the 4-meter Isaac Newton Telescope in the Canary Islands, at the same observatory as the much smaller 1.5-meter telescope that the Danes had used. With the Internet reasonably well established, the Berkeley group realized that the best way to process large amounts of data was to send it to its home computer in real time, so that an observing run in the Canaries involved not only astronomers at the telescope, but also Perlmutter, nine time zones away, monitoring the computer at the LBL.

In the spring of 1992, this approach yielded its first success, with a supernova discovered at a redshift of 0.458. By this time, Perlmutter and his colleagues had adopted the strategy that the Danish astronomers had introduced to optimize their chances of finding distant supernovae and exploiting their discoveries. Like much of astronomy, this strategy differed from almost all other human activity (despite what old wives' tales say about lunacy) in being directly tied to the changing phases of the moon.

THE INFLUENCE OF MOONLIGHT ON ASTRONOMERS

Astronomers who use giant telescopes to find supernovae must respect one of the most obvious factors limiting their perceptual capabilities: the light of the moon. To this day, and indeed for the foreseeable future, every major observatory must assign its observing runs—the nights granted to astronomers who have applied to use the telescopes—in an alternating cycle of "dark time" and "bright time." Dark-time observing runs include the two-week period centered on the day of a new moon, when the moon passes almost directly between the sun and the Earth. On that day, when the moon's entire lit side faces away from our planet and the moon lies below the horizon throughout the night, the nighttime skies are at their darkest. On the other nights during a dark run, the crescent moon either sets relatively early in the evening as it waxes toward first quarter or rises only near dawn as it wanes toward the new moon. So long as the moon is up only briefly or not at all, astronomers can observe faint objects through most of the night, unhindered by the moonlight that Earth's atmosphere scatters all around the sky, depriving astronomers of the ability to make precise observations of the faintest objects. The other half of the month, the bright moonlit nights, can be devoted to observations of relatively bright objects, often to spectral measurements of objects whose light output can compete successfully with the stray light from the moon.

Because only the dark time allows astronomers the chance to study the faintest objects with high precision, large observatories assign their "dark runs" to observational programs that concentrate on securing images of faint objects, on observing their spectra, or on measuring their brightnesses. High-redshift supernovae fall definitively in the category of faint objects, so all efforts to discover and to measure supernovae in far-distant galaxies must respond to the phases of the moon as surely as do the grunion that run through the moon-raised tidal waters of southern California.

Following the lead of the Danish supernova searches, the Berkeley group developed the highly successful tactic of timing its supernova searches to the 29.5-day cycle of the moon's phases. The astronomers decided to obtain an image of a selected area of the sky, containing many thousand galaxies visible to their telescope, during the second half of a dark run, just after a new moon. Then, about three weeks later, with the next dark run under way, they would secure a new image of the same galaxies. Thanks to modern computer techniques, the astronomers would require only a few hours to compare their new images with those taken three weeks before, to search for new bright objects—stars caught in explosion on the second observing run. In fact, this technique would allow the astronomers not only to detect a new supernova, but also to find it before it reached its maximum brightness. Each new supernova would naturally receive special attention from the astronomers, who would have at least a week remaining in the dark run to make careful measurements of the supernova's light output and spectrum.

This technique, which the Berkeley group named the "batch process," worked like a charm. After overcoming a few glitches, the group soon showed that if it were granted two batches of observing time—one at the end of a dark run and the other near the beginning of the next dark run—it could successfully predict the approximate number of supernovae it would discover. The success of the batch-process approach made it so obviously the road to success that the competing group of observers would adopt an essentially identical observing style.

One aspect of observing high-redshift supernovae slowed things down without causing serious difficulty. Far-distant galaxies spread over only a tiny angle on the sky: They look small because they are far away. When a supernova appears in one of these galaxies, astronomers cannot easily distinguish the brightness of the galaxy from the brightness of the supernova itself. As a result, their brightness measurements refer to the combined light of the galaxy and the supernova. The solution lies in waiting for the best part of a year, until the supernova has faded, and then reobserving the galaxy's brightness. Because galaxies do not change their brightness on timescales measured in years, the latter observation can be subtracted from the former to determine how bright the supernova must have been.

FIGURE 9.1   The Batch-Process Method for Discovering
Distant Supernovae

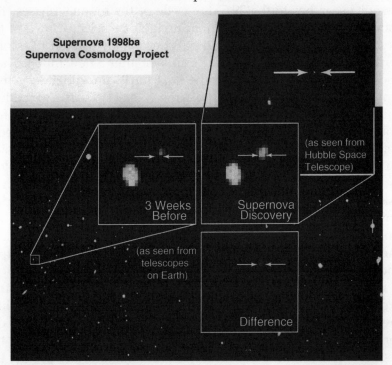

This schematic diagram shows the "batch-process" method developed by the Supernova Cosmology Project to discover and to follow large numbers of distant supernovae. The inset images show one of these galaxies three weeks before the discovery of a supernova within it and then at the time of discovery, with a blowup showing the supernova as photographed by the Hubble Space Telescope. This supernova outshines its host galaxy, which appears to the upper right of the supernova itself. (Courtesy of Dr. Saul Perlmutter et al. (the Supernova Cosmology Project).)

One exception exists to this rule that requires astronomers to wait for a year to complete their observations of a supernova: the Hubble Space Telescope, which professional astronomers familiarly call the "HST" (but almost never "the Hubble," because astronomers, who know the story of Edwin Hubble well, find this an odd confusion of the man with his monument, much like calling the John F. Kennedy Airport "the Kennedy"). The HST can obtain the crispest images available because it orbits above the Earth's atmosphere, which inevitably scatters and refracts some of the light passing through it, blurring astronomers' vision of fine details. Even though its mirror diameter of 2.4 meters would mark the HST as a puny rival to the giant optical telescopes on Earth's surface, the fact that it operates free from the blurring produced by our atmosphere allows it to defeat its larger brothers in the clarity of its observations.

The HST can even distinguish a supernova in a distant galaxy from the remainder of that galaxy, making it much easier for astronomers to measure the changing brightness of the supernova as it first climbs to a peak brightness and then fades into complete invisibility. In the days when organized supernova-detection programs remained in their infancy, no astronomer could have persuaded the HST's Time Allocation Committee to devote precious hours of observing time to galaxies that might prove to contain a newfound exploding star. The successes of the supernova observations, however, have shifted astronomers' attitudes. Supernova observers can guarantee that they will discover a significant number of new supernovae within a particular area of the sky long before they make their first observations—an impressive ability that has led them to be granted observing time with the HST. But all this occurred only after the experts had demonstrated that Type Ia supernovae can provide the standard candles for which astronomers had longed.

Mark Phillips Makes
the Key Discovery About Type Ia Supernovae

In 1993, as the Berkeley supernova group prepared to reap the first rewards from its batch process, an American astronomer named Mark Phillips, working at the Cerro Tololo Inter-American Observatory in Chile (supported in large part by the NSF), published a research paper whose results promoted Type Ia supernovae as the highly luminous standard candles that would unlock the secrets of the universe. Using a series of other astronomers' observations of the relatively nearby SN Ia's that had appeared in galaxies with reasonably well-estimated distances from the Milky Way, Phillips demonstrated beyond reasonable doubt that Type Ia supernovae exhibit a correlation between their peak luminosities and the rates at which their brightnesses decline: *More luminous Type Ia's fade more slowly than less luminous supernovae.*

A Soviet astronomer had noticed this effect two decades earlier, but astronomers had considered the data unreliable. Once Phillips had established the relationship between the peak luminosities and rates of decline in SN Ia's as trustworthy, Perlmutter and his colleagues seized on it as the key to using SN Ia's as standard candles. They soon derived a "stretch factor" that described how rapidly a particular supernova's light curve declines toward obscurity. Each value of this stretch factor corresponds to a particular value for the peak luminosity of a supernova, so that the tribe of SN Ia's could be separated into individual families, each attaining a slightly different peak luminosity and each identifiable by its stretch factor—that is, by the rate at which its apparent brightness decreased after reaching its peak. Before Phillips discovered this relationship, astronomers had spoken of their hope that SN Ia's could serve as standard candles, or of indications that they might do so. For example, Robert Kirshner, a highly respected supernova expert at Harvard, had reacted to the Danish discovery of the first high-redshift supernova of Type Ia by noting that "supernovae might lead us out of an age of [cosmological] ignorance and belief into an era of measurement and understanding," but he cautioned that much research remained to be done to verify the assumption that distant and nearby SN Ia's share identical properties.

THE BOYS FROM HARVARD

Mark Phillips's discovery changed astronomers' assessment of the usefulness of Type Ia supernovae as distance indicators. The Phillips relationship between each Type Ia supernova's peak luminosity and the rate of decline in its brightness made SN Ia's stand out, in all senses of the term, as the great new hope for determining the key cosmological parameters through observations of exploding stars in distant galaxies. By early 1994, as the new era of understanding SN Ia's got well under way, the Berkeley supernova group found itself engaged in a competition with a formidable group of astronomers, originally centered at the self-proclaimed mightiest of all universities and initiated by Kirshner in collaboration with Brian Schmidt, a Harvard-trained astronomer now working in Australia, who has assumed the leadership of the group.

This second group of supernova observers, officially named the "High-Z Supernova Search Team," had the slower start, but with hard and inspired work, it eventually caught up with Saul Perlmutter and the Supernova Cosmology Project. Unlike the Berkeley team, the competing group of supernova experts consists mainly of astronomers, who have little experience with, and less enjoyment of, hierarchical settings. In a movie, this fact would qualify the second bunch as heroes; in real life, the two groups have achieved roughly equal success and consist of similar sorts of individuals.

The second group lost all hope of securing the underdog label by centering its initial existence at that least underdoggy of American institutions, Harvard University. Harvard, a leading light in astronomical research for centuries, has recovered nicely from its early-twentieth-century lull, first by hiring Harlow Shapley and then by appointing dozens of the world's leading astronomers to its faculty. With the largest endowment of all universities, Harvard committees that seek new professors traditionally ask, Who is the world's leading scholar in this field, and what must we do to hire him? (Attempts to replace "him" with "her" have produced fights at Harvard at least as bitter as those at other universities; women now constitute 12 percent of the tenured Harvard faculty. Cecilia Payne, the greatest woman astronomer of the twentieth century, became a Harvard professor only after thirty years of demonstrating her abilities.)

Among the stellar talent assembled at Harvard's Center for Astrophysics, none ranks higher than Robert Kirshner, a professor of astronomy and former chair of the Astronomy Department. Now in his early fifties, Kirshner stands at the peak of his profession, known around the world as an expert in observing and interpreting supernovae. A Harvard undergraduate like Perlmutter, Kirshner enrolled at the California Institute of Technology for his graduate study in astronomy. Assigned an office in the second subbasement of the astronomy building, Kirshner often had the opportunity to converse with, and admire the soaring scientific imagination of, his office neighbor Fritz Zwicky, known to some at Caltech as "the mad Swiss," the man who had first conceived a connection between supernovae and the collapse of stellar cores.

Kirshner had chosen supernovae as the research subject for his thesis, mainly because he had written a paper as an undergraduate on the most famous remnant of a supernova, the Crab Nebula, and he therefore had something to say when his graduate thesis advisor asked the stern question, "What do you want to study?" Over the course of the decade after finishing his thesis, which dealt with observations of supernovae and their remnants, Kirshner continued this line of research, which led him to a faculty appointment at the University of Michigan and, in 1985, to a Harvard professorship. In February 1987, when a supernova appeared in the Large Magellanic Cloud, the Milky Way's satellite galaxy, Kirshner took a leading part in organizing and analyzing worldwide observations, including those made with the *International Ultraviolet Explorer*, a satellite capable of recording some of the short-wavelength electromagnetic radiation that cannot penetrate Earth's atmosphere. This exploding star, the closest supernova to the solar system during the past three centuries, has the astronomical name "Supernova 1987A"—the first supernova found during the year 1987.

Supernova 1987A turned out not to belong to the class of Type Ia supernovae that would come to dominate the cosmology story by providing standard candles. Instead, Supernova 1987A belongs to the entirely different class of Type II supernovae, which arise when the cores of massive stars collapse. Before Type Ia supernovae proved so useful as standard candles, some astronomers attempted to make the Type II's perform essentially the same function. Because the different Type II supernovae reach quite different luminosities, their apparent brightnesses cannot play the direct role that they do for Type Ia's. Astronomers could, however, measure the speeds at which these supernovae eject their outer layers into space, revealed from observations of their spectra and a knowledge of the Doppler effect. By making accurate correlations of these ejection velocities with the observed changes in each supernova's apparent brightness, astronomers could deduce the distances to Type II supernovae. Although useful and promising, this method of estimating distances has not yet yielded distance determinations as accurate as those made with Type Ia supernovae. Furthermore, because Type II supernovae reach a significantly lower peak luminosity than the Type Ia's, they cannot be observed at such immense distances as those of the high-redshift SN Ia's.

Kirshner, not the sort of person to focus on a single type of supernova, remained well acquainted with the various possibilities for using exploding stars to determine the distances to faraway galaxies. In addition to his university duties, he served on committees to review outside projects, including the automated supernova search that the Berkeley supernova group was developing. Kirshner recognized this method as full of promise, and he had an astronomer's suspicion that physicists inevitably would miss something important in obtaining, refining, and analyzing their data. He also knew that a competition between observing projects would offer an increased chance for useful results, not only from the adrenaline rush of beating the competition, but also from the chance for each group to explore its suspicions and check on the other's interpretations.

Early in 1994, Kirshner and Brian Schmidt, who had just finished his Ph.D. thesis, under Kirshner's supervision, dealing with Type II supernovae, created a group at Harvard to explore the use of Type Ia supernovae as standard candles. Kirshner, who is excellent at delegating authority to younger colleagues who can make the most of it, encouraged Schmidt to contact other supernova observers to help them gather data on Type Ia's. Quite understandably, the first members of this group were graduate students already inclined to work with Kirshner and Schmidt on supernova issues—most notably, Peter Garnavich, who had received his degree at the end of 1991, and Adam Riess, who became a Harvard graduate student in the fall of 1992. Bruno Leibundgut, an astronomer at the European Southern Observatory, headquartered in Munich, also joined the group, of which Schmidt became the designated leader in 1995, when the astronomers filed their first application to use the HST for supernova observations.

Schmidt's role in this team centered on designing and improving the software that would search the telescopic images for signs of supernovae, a task that had to be performed quickly, so that the supernovae could be followed as early as possible along their light curves. Schmidt, married to a graduate student from Australia whose visa called for her to leave the United States in 1994, felt himself fortunate to obtain a position at the Mount Stromlo Observatory near Canberra, but the state of the Internet in 1995 made his work a nightmare: To transmit a single image from Chile to Australia took forty-eight hours! Schmidt eventually overcame this ob-

stacle and developed a software system that could perform as well
as the one used by the Berkeley supernova group. Starting in late
1994, a couple of years behind the team at Berkeley, the competing
astronomers in Chile and Australia and at Harvard rapidly caught
up with their rivals in detecting and observing supernovae.

Meanwhile, in the offices of Harvard's Center for Astrophysics,
Adam Riess was developing new ways to interpret supernova
light curves. Riess, an extremely pleasant young man with the
mien of a scientifically oriented chipmunk, reminds some who
meet him of how Jerry Mathers, the portrayer of the Beaver of tele-
vision fame, might have looked after obtaining a Ph.D. in astronomy. This mien, however, has a deceptive power. Abraham
Lincoln's law partner, William Herndon, said that "the man who
thinks that Lincoln had no ambition is making a grave mistake.
Lincoln's ambition was a little engine that knew no rest." As was
true for Shapley, Hubble, Zwicky, Einstein, and Alvarez, and as is
true for Muller, Perlmutter, Kirshner, and Schmidt, the core of
Riess's personality lies in a drive to ferret out the secrets of the
universe.

Born in New Jersey to parents who encouraged his interest in science, Riess graduated in 1992 from the Massachusetts Institute of
Technology, often known at Harvard as "the science school down
the river," since MIT also lies in Cambridge, facing Boston across
the Charles River, which has there grown considerably wider than
the stream that separates Harvard's dormitories from its football
stadium and business school. As a physics undergraduate at MIT,
Riess had applied to the graduate program in astrophysics at Harvard without much knowledge of astronomy. By his second semes-
ter, when graduate students were expected to name a research
topic, Riess went looking for one. He spoke with Irwin Shapiro, the
head of the astrophysics program, who told him that if he (Riess)
chose to work with him, he (Shapiro) was too busy to check his results—so he had better be right. A bit stunned by this attitude, Riess
soon found Kirshner, probably the most noticeable astronomer at
Harvard, who charmed him with his humor and verve, and listed a
few projects that might serve their mutual purposes. By this time,
early in 1993, the use of Type Ia supernovae as standard candles
was much on Kirshner's mind, if not at all on Riess's. Kirshner
threw Riess a challenge, stating that some astronomers had suggested that Type Ia's could not serve this function.

Like any good scientist, Kirshner was eager to check whether a good idea could survive all attempts at refutation. He explained to Riess that the relatively few Type Ia supernovae with low redshifts appeared to display a wide range of peak luminosities. Some of this could be attributed to the absorption of starlight by interstellar dust in the Milky Way, but some of it must be intrinsic to the supernovae. What could Riess do about this? Could he find a way to determine the intrinsic luminosity of each Type Ia supernova, in which case the supernova's observed maximum brightness would reveal its distance?

Of course, Kirshner and Riess carefully examined Mark Phillips's recently published paper describing seven SN Ia's, which demonstrated the correlation between peak luminosity and the rate of decline of the supernovae. Could a more accurate method exist to relate each supernova's peak luminosity to its light curve, the time history of its changing apparent brightness? Riess found that there was. From 1994 through 1996, Riess became the data expert for the Harvard supernova group, an expert not at using telescopes, but at teasing the meaning from the data that other members of the group acquired. Using statistical analysis pioneered by Kirshner's colleague William Press, Riess not only transformed the raw observational data into the most accurate light curves attainable but also sought new ways to find what these light curves implied.

This analysis showed that the details of the light curves of Type Ia supernovae, which include but go beyond the parameter that describes how rapidly each supernova's brightness declines, can be matched with the peak luminosities of the different Type Ia supernovae. In a comparison of graphs that became famous within the supernova community, Riess showed that this "light-curve shape analysis" would allow astronomers to deduce a supernova's peak luminosity, and thus its distance, from the details of its changing brightness. In one graph, Riess plotted the absolute light curves for various supernovae—their deduced luminosities as a function of time—to show that these curves spread over a fairly sizable region on the graph. Then, in a second graph, Riess redrew the light curves, correcting them upward or downward in luminosity on the basis of his light-curve shape analysis. The spread nearly disappeared as the light curves now lay almost atop one another.

In 1996, when Riess and his colleagues published these results, they convinced most of the astronomical world that Type Ia supernovae could furnish the long-desired tool: standard candles visible at distances of billions of light-years. Like all good science, this conclusion did not pass unargued—a fact that emerged in spades when the news from Type Ia's dominated the world of cosmology. In 1999, the Astronomical Society of the Pacific awarded Riess its Trumpler Prize, given for the Ph.D. thesis within the past three years that has had the most significant impact on astronomical research.

As Riess's conclusions emerged, the Berkeley supernova group naturally questioned whether its somewhat simpler approach, relying on a single stretch factor in the light curve, could yield results as accurate as the light-curve shape analysis. The answer turned out to be basically yes. In fact, the full analysis of the Berkeley scientists' light curves revealed that a large majority of all of the distant Type Ia supernovae follow nearly the same history in their changing luminosities: They reach the same peak luminosity, display the same stretch factor, and exhibit the same shape of their light curves. Naively, this might seem to make Riess's (and Phillips's) work seem almost unimportant. This conclusion would be dead wrong. The happy summary—that most high-redshift SN Ia's behave almost identically—could be verified only by performing a detailed study of the relationship between the peak luminosities and the history of changing brightness of a significant number of Type Ia supernovae. Furthermore, the minority of Type Ia's that do not conform to the simplest summary had to be identified and dealt with before Type Ia supernovae could reveal the likely existence of a cosmological constant.

Before this occurred, Riess had submitted his Harvard Ph.D. thesis and had been offered a position with the Supernova Cosmology Project by Saul Perlmutter. Riess did move to Berkeley, but instead of joining Perlmutter's group, he became a research fellow in the Astronomy Department, working with the department's supernova expert, Alex Filippenko. Filippenko, now just past forty and therefore Perlmutter's coeval, has a wide grin and the happily engaged aspect of a hardworking American—the sort who has grown up in a bilingual household oriented toward learning, has graduated from high school at sixteen, has lived at home while distinguishing himself academically, and has become a standout

FIGURE 9.2 Adjustment for Light-Curve Shapes of Supernovae

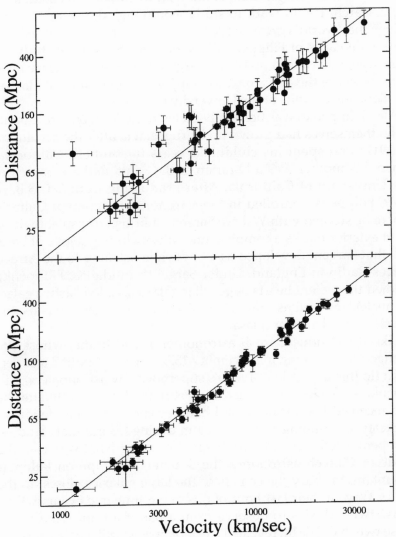

This figure shows the Hubble diagram derived from observations of Type Ia supernovae, before (top) and after (bottom) the peak brightnesses of the supernovae have been adjusted to conform to what relatively nearby Type Ia's have revealed about the correlation between their light-curve shapes and peak luminosities. The adjustment leads to a much tighter velocity-distance relationship. (Diagram courtesy of Dr. Adam Riess and the High-Z Supernova Search Team.)

graduate student at Caltech and a professor at Berkeley, where he wins awards for his teaching along with worldwide distinction for his research. If you want to study astronomy, the company that brings the "world's greatest teachers" into your home will sell you forty videotapes of Filippenko's lectures; if you want to study supernovae, you can do no better—arguably not even at Harvard—than to secure the office next to Filippenko's in Campbell Hall on the Berkeley campus of the University of California.

Alex Filippenko was born in Berkeley to Russian émigré parents (who themselves had grown up in Yugoslavia after the revolution of 1917) and spent his childhood years in southern California, where his mother was a librarian at the Santa Barbara branch of the University of California. After graduating from U.C.S.B. in 1979, Filippenko enrolled in the astronomy program at Caltech, where he studied with Wallace Sargent, a British transplant whose gruff exterior masks a complex inner life, including a love of baseball that Sargent developed as a boy while listening to the Armed Forces Radio in England. Under Sargent's guidance, Filippenko studied the violent hearts of peculiar galaxies, called "active galactic nuclei," as well as these objects' still more violent cousins, the quasi-stellar objects or quasars.

Like the famous Danish astronomer Tycho Brahe, whose life changed one evening in November 1572 when he spotted a supernova (to this day called "Tycho's supernova" by astronomers), Filippenko's life altered one night in February 1985, shortly after he had received his doctorate and had become a graduate fellow at Berkeley. Continuing a project begun during his graduate studies, Filippenko collaborated with Sargent in using what was then the pride of Caltech astronomy, the 5-meter telescope on Palomar Mountain, to study the centers of the large galaxies closest to the Milky Way. In their last hour of observing time one evening, they chose two galaxies to observe from a list of a hundred. One of these two, NGC 4618, revealed a bright, new starlike object near its center. The spectrum of light from this object was unlike any that Filippenko had seen, and it became even more unusual as the weeks passed. The exploding star in NGC 4618 was the first representative of the Type Ib supernovae to be observed long after its maximum light output. Type Ib supernovae have spectra that show no evidence of hydrogen but do reveal features arising from the absorption of light by helium. From that time forward, Filippenko changed the main focus of his astronomical attention from active galactic nuclei and quasars to supernovae. His efforts, along with others, helped astronomers to single out Type Ia supernovae as the special objects whose explosions are described in the next chapter.

Even before the discovery that flipped him from galaxies to supernovae, Filippenko had been awarded a prestigious Miller Fellowship by the University of California at Berkeley, which provided him with support for two years of scientific research unhampered by the teaching duties for which he would later win awards. Filippenko became a leading expert on the interpretation of the spectra of light from exploding stars and, in particular, on Type Ia supernovae; before long, the university made him an associate and then a full professor. As an observationally oriented astronomer, he treasured the opportunity to use the world's largest telescope, the 10-meter Keck reflector on the peak of the Mauna Kea volcano in Hawaii (first fully operational in 1992 and now joined on the summit by its twin, the Keck II). Jointly operated by Caltech and the University of California, which themselves have now joined a partnership with NASA (which has the greatest interest in using the telescopes to find planets around other stars), the Keck twins combine the greatest light-gathering power of any optical telescopes with the finest site, nearly 14,000 feet above sea level, a crown of calm above most of the atmospheric turbulence, which floats by at lower altitudes. Each Keck telescope can secure spectra with a precision unmatched the world over—just what is needed to determine whether a faint, distant object is a supernova and (even more important), if so, what type of supernova.

Early in his professorship, Filippenko had an offer to join the Caltech faculty, but he chose to remain in Berkeley, influenced by his love for the state university system, as well as by the opinion of his then fiancée, a graduate of U.C. Berkeley's law school, who has a familiar northern California attitude toward Pasadena and the rest of the Los Angeles basin. Caltech thus lost its chance to rectify a strange situation: Arguably the world's greatest collection of scientists pound for pound, the place where scientists first began to understand supernovae, Caltech has no true supernova expert, while Harvard and Berkeley carry the supernova torch to ever-higher altitudes.

In the early 1990s, Perlmutter persuaded Filippenko to become a member of his supernova group, which was noticeably top-heavy in physicists and needed an astronomer who understood supernova spectra. Filippenko contributed to the group's early successes, but early in 1996, he switched over to the competing High-Z Supernova Search Team. "I don't like to sit around juggling balls that have only minor effect," Filippenko notes of his disjunction from Perlmutter's group. The Harvard group's much looser organization, typical of how astronomers interact, was far more congenial to Filippenko, who helped secure a Miller Fellowship for Riess to join him at U.C. Berkeley's Astronomy Department late in 1996. Three years later, Riess left Berkeley to become a staff astronomer at the Space Telescope Science Institute in Baltimore, Maryland. By now, with Filippenko at Berkeley, Riess in Baltimore, Brian Schmidt in Australia, and Mark Phillips and Mario Hamuy in Chile, the name "Harvard group" has become a misnomer that raises the hackles of non-Harvard astronomers, though Kirshner and his former Ph.D. students Riess and Schmidt do play crucial roles in this group's efforts.

The shorthand "Berkeley group" for the Supernova Cosmology Project can likewise cause confusion: Both groups of supernova experts have created a significant presence in Berkeley, perhaps the most famous small town in the United States. Such fame was never more deserved than in 1997 and 1998, when supernova astronomers led cosmology onto a path few had anticipated. The first announcement of observational results from high-redshift supernovae came from Saul Perlmutter's group of astronomers, who published in mid-1997 what they had found from the first seven high-redshift supernovae they had studied. These results startled the astronomical community because they implied a strikingly large value for Ω_M, perhaps as large as 1, while implying that the cosmological constant should be near 0. Throughout the last months of 1997, these supernova results produced a high level of speculation and interest. Everyone knew, however, that seven high-redshift supernovae were too few to avoid the pitfalls of being misled by one or two rogue objects. And so it turned out: Examination of the results from a larger set of high-redshift supernovae revealed quite a different, even more startling, result.

In January 1998, the Supernova Cosmology Project announced its new results at the meeting of the American Astronomical Soci-

FIGURE 9.3   The Hubble Diagram,
Including Forty-two Distant Supernovae

Supernova Cosmology Project

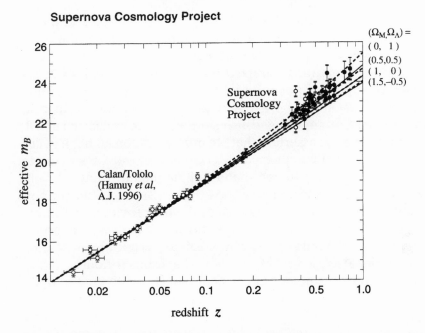

Presented here is the velocity-distance diagram obtained by the Supernova Cosmology Project, including the group's observations of forty-two Type Ia supernovae with redshifts of 0.4 or more. The low-redshift portion of the diagram
shows data from observations of relatively nearby supernovae in what astronomers call the "Calan/Tololo survey." The diagram's vertical axis denotes
distances by using an astronomical term called the "effective blue magnitude"
(m_B), a measure of apparent brightness, of the supernovae at maximum light.
The different lines at the top right show the predictions of different model universes, specified by the values of Ω_M and Ω_Λ. All the models shown have Ω_M +
Ω_Λ = 1. These results are in overall agreement with those of the High-Z Supernova Search Team, shown in Figure 7.1. (Diagram courtesy of Dr. Saul Perlmutter et al. (the Supernova Cosmology Project).)

ety in Washington, D.C. These received modest worldwide attention, but Perlmutter and his colleagues chose not to emphasize just
what their work product implied because significant uncertainties
remained in their data analysis. The publicity wave broke in February, when Alex Filippenko stood before a conference of astro-

physicists in southern California and announced that his group's analysis definitely implied that the cosmological constant can no longer be taken as zero.

In Chapter 7, we met the reasons for this conclusion and the expansive effects that a cosmological constant has on the evolution of the universe. Nevertheless, all those not well acquainted with the supernova observations greeted the announcements by the two groups of supernova experts with deep skepticism. Something, they said, must surely be wrong with the observations or their interpretation. Certainly this conclusion has a far greater chance to be correct than the assertion that we must abandon all hope of cosmic recycling and accept the existence of a nonzero cosmological constant. In Chapter 10, we shall examine the most important attacks on the results from the observations of supernovae. For now, we should note that if a single team of supernova observers had announced what the two groups did in January and February 1998, the world would have rightly demanded additional scrutiny and additional evidence before taking the results seriously.

The nonzero cosmological constant gained relatively rapid and widespread acceptance precisely because two teams of observers, not much enamored of each other and deeply suspicious that the other group had missed something important in obtaining and analyzing the data, had reached the same conclusions from independent sets of data. No one could claim more experienced or deep-seated skepticism than either of these rivals in examining the other's efforts. From this situation came the new cosmology, in which the cosmological constant must apparently be admitted as a full member of the parametric family, one of the crucial numbers that rules the universe.

An Intriguing Sidelight: Supernovae Demonstrate the Slowing Down of Time

In the preceding discussion of Type Ia supernovae, including that of the relationship between the peak luminosities of Type Ia supernovae and the rates at which their brightnesses decline, we have passed over astronomers' adjustment of the supernova light curves to allow for the slowing down of time. This effect, predicted by Einstein's special theory of relativity, arises when we observe a

FIGURE 9.4   Results for Ω_M and Ω_Λ Obtained by
the Supernova Cosmology Project

Supernova Cosmology Project

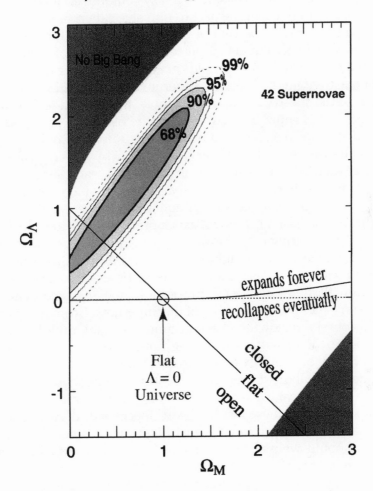

The data obtained by the Supernova Cosmology Project strongly imply that Ω_Λ must be greater than 0 and that Ω_M and Ω_Λ may well sum to 1. The likelihood contours show the region of the diagram within which the values of Ω_M and Ω_Λ are likely to reside, with confidence levels equal to 68 percent, 90 percent, 95 percent, and 99 percent. Like the data obtained by the High-Z Supernova Search Team, shown in Figure 7.3, these results exclude a universe in which $\Omega_M = 1$ and $\Omega_\Lambda = 0$. (Diagram courtesy of Dr. Saul Perlmutter et al. (the Supernova Cosmology Project).)

system of particles—a supernova, for example—that is moving at a high velocity with respect to us. We observe that time in the moving system passes more slowly than it does in our reference system. The slowing down of time becomes ever more noticeable for progressively larger velocities. If a system moves at 60 percent of the speed of light with respect to ourselves, then time passes in that system at only 80 percent of the rate that it does for us; and if the relative velocity equals 80 percent of the speed of light, then time in the moving system unfolds at 60 percent of our rate.[1]

When astronomers observe a supernova receding from us at 60 percent of the speed of light, for example, they expect to find that the supernova's entire light curve, from the rise to the peak and on through slow decline, shows a history that unfolds at only 80 percent of the rate that they observe for a nearby supernova with identical qualities. As discussed in Chapter 4, each supernova's redshift reveals its recession velocity. A redshift of 2, for example, corresponds to a recession velocity equal to 60 percent of the speed of light. For such a supernova, astronomers expect to observe a light curve that embodies the slowing down of time, with the result that the supernova's light curve passes through its different stages only 80 percent as rapidly as the light curve for a nearby supernova with identical spectral properties. First analyzed in detail by Gerson Goldhaber, a member of the Supernova Cosmology Project, this observed slowing down of time in distant, rapidly receding supernovae provides one of the finest proofs that Einstein's special theory of relativity enjoys cosmic validity.[2]

Speaking of validity brings us to the most fundamental scientific reaction to the breaking news in cosmology: How much credence should we place in these results? More specifically, do they rise to the level that everyone must adjust his or her cosmological attitude, or should we maintain a discreet reserve, awaiting further observations that will confirm or refute them? This question deserves a chapter to itself, of which we can now furnish a summary: As is true for many great questions in science, the verdict on the value of the cosmological constant awaits the future.

---

[1]The rates at which time passes in the two systems are mathematically linked by the square root of the factor $[1 - (v/c)^2]$.

[2]Strictly speaking, the recession velocity, which arises from the expansion of the universe, brings into play Einstein's general theory of relativity as well as his special theory. For the purposes of this book, however, we lose little in concluding that the supernova observations help validate the special theory.

CHAPTER TEN

COULD THE COSMOLOGICAL NEWS BE WRONG?

SCIENTISTS HAVE LONG KNOWN THAT the announcement of any significant result must receive and pass the best attacks on its every facet before its gemlike qualities can stand adamantly before the world, announcing that humans have discovered another slice of nature's truth. We may now apply this principle to the news that ranks as the greatest cosmological discovery of the century's final years—provided that the results prove to be correct.

A SALUTARY HISTORY: DISTANCE DETERMINATIONS BASED ON CEPHEID VARIABLE STARS

Astronomers have achieved only a limited understanding of how Type Ia supernovae explode. This fact leaves the astronomical community in much the same position, so far as the cosmological implications of SN Ia observations are concerned, that it occupied three-quarters of a century ago in its quest to determine the distances to spiral nebulae. In that earlier era, Cepheid variable stars provided good standard candles, although three decades would elapse before astronomers understood the physical processes that govern the Cepheids.

Astronomers such as Henrietta Swan Leavitt could recognize Cepheid variables as good standard candles because a sizable number of them existed at roughly the same distance from the solar system, in the Small Magellanic Cloud. At an equal distance from us, the Cepheids' apparent brightnesses corresponded directly to their luminosities, which allowed Leavitt to recognize that the variable stars with the longer periods of light variation have the greater luminosities. The next and crucial step, taken by Edwin Hubble and promptly adopted by other astronomers, was to assume that any Cepheid variables seen in spiral nebulae had characteristics identical to those of Cepheids in the Small Magellanic Cloud and our Milky Way. In particular, astronomers assumed that the relationship between a variable star's luminosity and its period of light variation remained unchanged as they shifted their gaze from relatively close to more distant Cepheids.

This assumption seemed justified by what astronomers call the "cosmological" or "Copernican principle," the notion that Earth occupies no special place in the cosmos—by extension, neither does our sun, our solar system, or our Milky Way galaxy. This principle governs astronomical concepts of the universe; to reject it would amount, in astronomers' eyes, to special pleading on behalf of our situation in space. (In Chapter 14, we shall confront the pain of rejecting the principle in the flow of time rather than in the extent of space.) So far as astronomers can tell, Cepheid variable stars in the Milky Way, in the Small Magellanic Cloud, and in the Andromeda and other spiral galaxies have nearly identical properties, so long as we can identify them as belonging to the same class of objects and to the same category within that class. Despite the validity of the Copernican principle, astronomers made serious errors for nearly thirty years in using the standard-candle method of comparison to estimate the distances to galaxies. This happened because they mistakenly confused what turned out to be two separate categories of Cepheid variable stars in their attempts to determine the distances to Cepheids within our own galaxy.

No Cepheid variable star lies sufficiently close to the solar system for its distance to be measured accurately with the parallax effect. Cepheids are unusual, highly luminous objects that arise during the late stages of a massive star's nuclear-fusing lifetime. The closest of them, we now know, lies at distances nearly a thousand times greater than those of the closest stars to the sun. How, then, could astronomers calibrate the distance scale of Cepheids, which then might extend far into space by observing fainter, more distant Cepheid variables?

The answer lay in long-term, accurate observations of the positions on the sky of Cepheid variables in the Milky Way. Over several decades, astronomers could detect small shifts in the stars' positions, which must arise from the stars' motions through space with respect to the motion of the solar system. The shifts on the sky, called "proper motions," reflected only one part of the stars' movements: the component of motion perpendicular to our line of sight to each star. The other component, which lies parallel to our line of sight, revealed itself in the stars' spectra, where the Doppler effect produces changes that reflect the amounts of any velocities of approach or recession.

For a large number of stars, astronomers reasonably expected that the velocity component along our line of sight should roughly equal the component perpendicular to the line of sight. The parallel component emerges from Doppler-shift measurements in kilometers per second. The perpendicular component of motion, however, appears as an angular displacement whose amount depends on the actual velocity across our line of sight divided by the star's distance. Thus, if astronomers assume that Cepheids' angular displacements arise from velocities across the line of sight that roughly equal the velocities along the line of sight, they can obtain the stars' distances by simple algebra.

During the 1930s and 1940s, astronomers used this technique to derive the distances of Cepheid variables in the Milky Way. Comparison of these Cepheids with others, seen in distant galaxies, then provided the scale of distances in the universe. Everything made sense—but, in fact, astronomers misestimated the distances to other galaxies by immense amounts!

Two Effects that Confused Astronomers' Distance Estimates

How could astronomers have gone so far wrong? The answer to this question bears directly on the trust that we may place in the announcement of a runaway universe. Two separate sources of error, both highly relevant to the determination of distances with supernova observations, had thrown a monkey wrench into astronomers' conceptual machinery. The first and less important of these error sources consists of the absorption of starlight by interstellar dust, which concentrates in great clouds of gas and dust within the disks of spiral galaxies such as our Milky Way. This absorption makes distant stars appear even dimmer than their greater distances imply, and it causes astronomers to overestimate their distances. Hubble and his successors recognized this effect and strove to correct for it by determining the amount of interstellar dust along a particular line of sight. This presents a challenging task, which we shall discuss below in the context of estimating the distances to supernovae.

The second source of error worked in the opposite direction, causing astronomers to underestimate the distances to galaxies. Cepheid variable stars turn out to come in two types, which arise in the evolution of stars that contain noticeably different fractions of the elements heavier than hydrogen and helium. One type of Cepheids, the "classical Cepheids," contain about the same fractional abundance of these heavier elements as the sun does. The other type of Cepheids, which appear primarily in globular star clusters, contain only about one-tenth as much of the heavier elements. As theoretical calculations eventually demonstrated, the different abundances of the heavier elements produce significant differences in the luminosities of the variable stars. In 1949, when Walter Baade used the newly inaugurated 5-meter telescope on Palomar Mountain to photograph Cepheid variables in the Andromeda galaxy, he demonstrated that a classical Cepheid with a twenty-four-hour period of light variation has a luminosity about four times greater than a globular-cluster Cepheid with the same period. By merging the two into a single, imaginary type of standard candle, astronomers had underestimated the distance to the Andromeda galaxy by comparing its classical Cepheids with globular-cluster Cepheids in the Milky Way.

In a single stroke, Baade doubled the scale of the universe. Astronomers' estimates of the distances to all other giant galaxies rest on the distance to the closest of them, the great spiral in Andromeda. Further, more modest corrections, arising from closer study of the globular-cluster Cepheids and other objects used to estimate distances, have raised the distances to galaxies beyond Andromeda still more than Baade's discovery did. Today astronomers estimate the distance to the Andromeda galaxy at 2.2

million light-years, more than twice the value that Edwin Hubble obtained in 1923. A nearby large cluster of galaxies, in the constellation Coma Berenices, lies 250 million light-years away, rather than the 45 million that Hubble and Humason estimated in 1931.

SYSTEMATIC ERRORS:
THE BANE OF FIRM CONCLUSIONS

What, as the playwright Bertolt Brecht might have asked, are the lessons learned from astronomers' errors in estimating the distances to galaxies? First, any determination of the distances to objects millions of light-years away represents fiendishly difficult work. How much more true must this judgment be when we look not millions, but billions of light-years beyond the Milky Way!

Second, astronomers often have more accurate values for distance *ratios* than for the absolute distances themselves. This occurs because the actual distances embrace a series of steps, including the calibration of the distance scale, that involve comparisons among different classes of objects. The distance ratios often involve only the comparison of what astronomers believe to be the same type of object. The story of the Cepheid variables, however, reminds us that astronomers face a huge problem in reaching near certainty that they are observing the same type of object at relatively near and far distances.

Third and most generally, astronomers must ceaselessly maintain alertness to possible sources of what they call "systematic error," such as the systematic differences between classical and globular-cluster Cepheids. Systematic errors differ from the errors that arise from inaccuracies in data collection, or from having only a few objects to analyze, some of which may prove highly anomalous once a larger data set emerges. Unlike the first source of error, which diminishes with increasingly accurate observations, or the second, which declines as astronomers observe more objects, systematic errors yield only to the insight that can recognize them, rather than to improved instruments or additional observations. Systematic errors are the bane of scientific conclusions based—as almost all conclusions are—on the comparison of two sets of results, such as the experimental and control groups in a medical test or the nearby and distant supernovae in a cosmological one.

When the supernova news burst on the astronomical scene in 1997 and 1998, the conservative reaction first focused on the obvious issues that arose from the difficulties in observing faint supernovae billions of light-years from Earth, as well as from the small number of distant supernovae whose characteristics had been studied. Indeed, the initial report from the Supernova Cosmology Project, based on the first seven supernovae under scrutiny, proved to be skewed by small-number statistics: These seven did not well represent the totality of distant supernovae. Once the astronomers in both the Supernova Cosmology Project and the High-Z Supernova Search Team had observed larger numbers of supernovae, and once they had carefully critiqued both their own and each other's results, probing for possible errors in observation, they and their peers concluded that the observations seem to have no statistical or observational flaws sufficiently serious to deflect the runaway universe. The greatest worry of all still remained: Do the supernova observations suffer from systematic effects that undermine the conclusion of an accelerating cosmos?

In 1989, commenting on the Danish discovery of the first high-redshift supernova, Robert Kirshner wrote that "[a]lthough it is tempting to think that supernovae might lead us out of an age of [cosmological] ignorance and belief into an era of measurement and understanding, two observational issues need to be carefully studied before too much faith is placed in this promising approach." Kirshner first raised the issue considered in Chapter 8: determining the extent to which Type Ia supernovae can serve as standard candles. Then Kirshner perceptively noted that "we need to build confidence that the supernovae observed at high redshift are really the same as the supernovae observed nearby" so that we might "be certain that any observed effect comes from space curvature [i.e., the effects of Ω_M and Ω_Λ] and not from a changing population of supernovae."

A changing population of supernovae would embody exactly the sort of systematic errors that astronomers fear in applying the standard-candle method to estimate distances. In raising the possibility that high-redshift SN Ia's might possess even slightly different characteristics from those of their low-redshift relatives, Kirshner put his finger on one crucial concern that astronomers would confront after assembling a statistically sufficient population of high-redshift supernovae. The other possibilities for systematic errors arise not with the supernovae themselves, but with the way in which we observe them. Could the cosmos have placed unperceived matter along the lines of sight to the high-redshift supernovae, matter whose effects lure astronomers into making false conclusions when they compare their observations of high-redshift and relatively nearby supernovae? Let us examine each of these two issues, walking a fine line between easy rejection of unknown systematic effects and the hagridden belief that we may never discover the truth about the universe.

DO SYSTEMATIC DIFFERENCES EXIST BETWEEN HIGH-REDSHIFT AND LOW-REDSHIFT SUPERNOVAE?

When astronomers observe supernovae with redshifts between 0.3 and 0.7, they are studying objects whose light has been traveling for immense intervals of time. These redshifts take us back to times when the universe had between 67 percent and 45 percent of its present age—further back, of course, for the supernovae with the larger redshifts. Thus these supernovae take us back four to seven billion years in time.[1] We must surely admit the possibility that the Type Ia supernovae exploding in those bygone eras differed, at least in subtle ways, from those that exploded only a few hundred million years ago, and therefore qualify as relatively nearby supernovae. For example, the production of heavy elements by supernovae, the explosions of which distribute the elements throughout the cosmos, provides newly formed stars that form with progressively larger fractions of these heavy elements as time goes on. Precisely this increase in heavier elements makes the luminosities of classical Cepheids significantly greater than those of globular-cluster Cepheids. This difference, which fooled astronomers for decades, amounts to a factor of four in luminosity. The existence of a nonzero cosmological constant rests on a much smaller luminosity factor, of about 25 percent. If "modern" SN Ia's turn out to reach luminosities 25 percent greater than those of the high-redshift supernovae, the cosmological constant will revert to zero, along with the reputations of many of the astronomers who have celebrated its glorious nonzerosity.

---

[1]Both the fractions of the universe's present age and the look-back times to the supernovae in question refer to a universe in which $\Omega_M + \Omega_\Lambda = 1$. If this sum has a value different from 1, the fractions and look-back times will deviate somewhat from those cited.

In the summer of 1999, a detailed comparison of the changes in the apparent brightnesses of relatively nearby and distant Type Ia supernovae seemed to reveal at least one systematic difference between the two supernova groups, thereby casting a shadow over the conclusions that these supernovae provide completely reliable standard candles. This difference appears in the "rise times" of the Type Ia's, the amount of time that passes between a supernova's initial explosion and the time that it achieves maximum luminosity.

Some uncertainty inevitably exists in determining these rise times, because we cannot hope to see the moment when a supernova explodes. Instead, each supernova reveals itself only at some time after its actual outburst, as its luminosity rapidly increases. The observed light curve of any supernova therefore begins at least a day, and more typically a few days, after the supernova explodes.[2] But the supernova's luminosity is then rising so rapidly that astronomers can rather easily extrapolate their observational record back toward the moment of explosion, introducing only modest errors in establishing this moment. Working with half a dozen colleagues, Riess, Filippenko, and Schmidt obtained observational data for ten relatively nearby Type Ia supernovae. One of these colleagues, Chuck Faranda, is a skilled amateur astronomer in Florida, whose 10-inch refracting telescope has an excellent CCD detector. Devoted amateurs such as Faranda need not apply to an allocation committee for telescope time; instead, they can search for supernovae as often as weather permits. In May 1998, Faranda discovered a Type Ia supernova, designated "SN1998bu," in the relatively nearby galaxy M96, only one day after its explosion and more than eighteen days before it achieved maximum luminosity—the earliest detection of any Type Ia supernova.

Once Riess and his colleagues had assembled their observational data for nearby Type Ia supernovae, they analyzed them statistically to derive an average rise time to maximum luminosity equal to 19.5 days. Then they turned to the sample of forty-two distant supernovae obtained by their rivals, the Supernova Cosmology Project headed by Perlmutter, the greater numbers of which offered the opportunity for a more accurate statistical analysis than

---

[2]We should always bear in mind that we see all these events millions or billions of years after they actually occurred, since the light they produce must travel for millions or billions of light-years to reach us.

would be possible with the smaller numbers of Type Ia's observed by the High-Z Supernova Search Team. To compare the data that describe the distant and nearby supernovae, the astronomers must allow for the slowing down of time described in the previous chapter, which makes the luminosities of distant supernovae seem to rise and fall more slowly than those of nearby supernovae. The measured redshifts in the spectra of distant supernovae allow astronomers to easily determine the speeds at which they are receding from us, and they also allow them to calculate how rapidly the luminosities of the distant supernovae would change if they had no significant velocity with respect to ourselves, as is true for the relatively nearby supernovae.

These allowances, easily made in accordance with the theory that Albert Einstein first created, in turn allowed Riess, Filippenko, Schmidt, and Weidong Li (a postdoctoral fellow in astronomy at U.C. Berkeley) to find that the distant Type Ia supernovae have a rise time to maximum luminosity close to 17 days—2.5 days less than the rise time for the nearby supernovae. If this discrepancy proves to be real, rather than a statistical artifact arising from the fact that only a relatively small number of supernovae are available for analysis, Riess and his colleagues will have discovered the first systematic difference between nearby and distant supernovae of Type Ia.

During the summer of 1999, the news of this systematic difference reverberated through the high-tension arena of cosmology, reminding all participants that once again, observations implying the existence of a cosmological constant might soon prove illusory. Two questions became paramount: What is the statistical significance of the difference discerned in the rise times for the two groups of Type Ia supernovae? And if this difference is real, what does it imply about the maximum luminosities that nearby and distant Type Ia's attain—that is, about their reliability as standard candles?

The second question has the shorter answer: No one knows—the most marvelous of all scientific possibilities. If astronomers possessed well-developed computer models of how Type Ia supernovae explode, they could vary the conditions in these models to determine which physical differences among supernovae will cause variations in their rise times. Then they could examine whether, and by how much, these differences affect the maximum luminosities that the supernovae reach. This would allow them to assess the robustness of their conclusion that, after the adjustments made on the basis of their light curves, all Type Ia supernovae, whether distant or nearby, reach nearly the same maximum luminosity. For now, lacking these detailed models, astronomers must speculate more broadly. Filippenko points out that if conditions near the surfaces of the white dwarfs vary systematically between the nearby and distant Type Ia supernovae, these differences might affect only the rapidity with which the explosion begins, with no significant effect on the maximum luminosity that the explosion produces. Assessing the situation with rough statistics, he judges it "a fifty-fifty chance" that the rise times, whether or not they differ systematically between nearby and distant supernovae, have nothing to do with the maximum luminosities of Type Ia's.

The other burning question deals with the statistical significance of the difference in rise times. Like almost all statistical issues, this analysis must follow a complex path. In theory, the significance of the rise-time difference can be derived from assessing the fraction of the difference between the two groups of supernovae that could arise from the fact that almost every Type Ia supernova has a rise time that differs somewhat from the mean value. Initially, Riess and his colleagues found that the discrepancy in rise times could almost certainly not arise from this fact but must instead represent a real and systematic difference between nearby and distant supernovae. However, it is possible that the quoted errors in the Supernova Cosmology Project's determination of the rise times for distant supernovae are overoptimistic. The members of the Supernova Cosmology Project have made their own analysis of these data and pronounce themselves unimpressed by claims of a systematic difference between the rise times of nearby and distant Type Ia supernovae. "I don't think that the [rise-time difference] has much statistical significance," Perlmutter says. "The data are simply not able to demonstrate a significant difference at this stage. Of course, we're all looking forward to obtaining better data."

This statement implies that Perlmutter shares the view that the rise times for the distant supernovae may not be known as accurately as the published data imply. The difficulty of accurately observing the changing brightnesses of distant, and therefore extremely faint, supernovae (especially as they begin their rise toward maximum light) may well have led to subtle errors, as yet undetected, that do not enter the statistical analysis. "I'd say the odds are fifty-fifty that [this sort of error] is the source of the apparent discrepancy," says Filippenko.

Thus the announcement that a systematic difference exists between the rise times of nearby and distant supernovae of Type Ia may prove to be another typical case in which a more careful analysis shows that the claimed phenomenon does not exist. On the other hand, this sequence of events points to just how small and subtle of an effect provoked the cosmological furor of the final years of the millennium. To summarize matters once again, the cosmological constant's claim to a nonzero value fundamentally rests on the finding that distant Type Ia supernovae reach maximum brightnesses approximately 25 percent fainter than the peak brightnesses they would attain in a universe with a cosmological constant equal to zero. If the conditions provoking these explosions varied in the past, with the result that Type Ia supernovae occurring billions of years ago reached maximum luminosities 25 percent less than similar explosions do today, we can wave goodbye to the cosmological constant.

Astronomers do have some arguments against this conclusion, based both on (admittedly incomplete) theories of supernova explosions and (indirectly) on observational data. On the theoretical side, astronomers know that a wide variation exists among the ages of white dwarfs, both now and at the times when the high-redshift supernovae exploded. White dwarfs can endure for billions of years and may accumulate matter from a companion star billions of years after becoming white dwarfs, as their companions, which are born with different masses and evolve at different rates, finally reach the stage in their lives at which they will transfer matter onto a nearby white dwarf. Therefore, if the peak luminosity of a Type Ia supernovae depended on the epoch in cosmic history when the white dwarf formed, we would expect to see a wide variation in the peak luminosities deduced for low-redshift supernovae, which arise from white dwarfs of widely disparate ages, even for those with the same stretch factor or shape of the light curve. Because supernova experts do not observe such a spread in luminosity, these experts conclude that the ages of white dwarfs that trigger SN Ia's probably do not affect the outcome and, in particular, leave the relationship between peak luminosity and light-curve shape undisturbed.

On the observational front, everything that astronomers can observe about Type Ia supernovae at low and high redshift supports the assumption that these exploding stars have nearly identical characteristics. These features include not only the light-curve shapes and stretch factors, but also the details of the peaks and valleys all through their spectra (adjusted, of course, for the Doppler effect) and at all times after the initial outburst. Because these spectra reflect the composition of the matter producing the light, or of matter absorbing light from regions closer to the center of the explosion, their near identity from near to far-distant supernovae shows that all Type Ia supernovae apparently have close to the same mixture of elements, as well as the same peak luminosities. Nevertheless, as Riess, Filippenko, and their eighteen coauthors wrote in their key paper, published in the *Astronomical Journal* in September 1998, "[a]lthough our current observations reveal no indication of evolution of SN Ia's at z [roughly equal to] 0.5, evolution remains a serious concern which can only be eased and perhaps understood by future studies."

Supernova evolution—changes in the characteristics of exploding stars as billions of years pass by—may therefore be placed on the top shelf of astronomers' concerns as they observe a public giddy with news of a nonzero cosmological constant. Just as high as evolution, and possibly deserving a shelf to itself, rests the anxiety that an effect extrinsic to any supernova might sow confusion in interpreting the observations of stars that exploded billions of light-years from Earth. That extrinsic effect has the astronomical title of interstellar and intergalactic absorption by dust grains.

THE PROBLEMS CAUSED WHEN
DUST PARTICLES ABSORB STARLIGHT

In analyzing the cosmic news brought to us by the light from distant supernovae, astronomers confront a difficult and subtle problem created by anything that absorbs some of the light on its way to us. We already know that whenever starlight passes through interstellar space in the Milky Way, dust grains absorb some of the light, misleading anyone who assumes that light travels without any such blockage.

Astronomers often refer to the process by which dust grains absorb starlight as interstellar "reddening," because the effectiveness of the dust grains in blocking the passage of starlight depends on the wavelength of the light: The grains absorb shorter-wavelength (violet and blue) light more effectively than they do longer-wavelength (orange and red) light. As a result, the light becomes redder, meaning that the preferential removal of violet in comparison with red has increased the ratio of red to violet light. In fact, both long-wavelength and short-wavelength light have been absorbed, but comparatively more of the short-wavelength radiation has. In the Earth's atmosphere, molecules and dust grains scatter sunlight in a manner similar to the effects of interstellar dust particles: They scatter blue light more efficiently than red light. This gives the sky its bluish hue and makes the sun look red near sunset and sunrise, when the beams of sunlight slant through more layers of Earth's atmosphere, exaggerating the normal effect.

The reddening caused by interstellar dust produces no beautiful sunsets. Instead, this reddening interferes with astronomers' attempts to deduce distances by comparing the apparent brightnesses of low-redshift and high-redshift supernovae. Because the expansion of the universe shifts what would be the blue light in a nearby supernova into the red region of the spectrum, whenever astronomers compare a nearby supernova's output in blue light with a distant supernova's brightness in red light, they inevitably contrast light that has experienced different amounts of absorption by interstellar dust. Interstellar dust resides mainly within galaxies, where the density of dust particles depends, in a general but imperfect way, on the distance from the galaxy's center. The interstellar reddening of any supernova's light therefore arises both from the dust within the Milky Way and within the "host galaxy," as astronomers call the galaxy containing the object under observation.

By now, astronomers have a fairly good idea of the spatial distribution of dust grains within our own galaxy, but they have only the most general notion of how the dust within the host galaxy has affected the light from a supernova when it began its journey. When they observe a high-redshift supernova, they cannot see fine details in the configuration of its host galaxy, and they can only estimate approximately the effects of dust particles there on the light from the supernova. Finally, interstellar absorption and reddening must occur throughout intergalactic space, since these regions cannot be completely devoid of dust. Fortunately, however, the density of dust particles between galaxies falls to such low values that we may probably neglect with safety any intergalactic effects in comparison with much larger, and largely unknown, effects that arise in the interstellar regions within galaxies.

Happily, astronomers have determined the correlation between the absorption of light produced by dust particles and the reddening that the dust causes. Reddening refers to preferential absorption—that is, to the difference in absorption of red and blue light. Unlike the absorption itself, the reddening of the light from a distant object reveals itself rather straightforwardly, by changing the ratios of the light of different colors from a familiar type of object. Although astronomers must make some fancy adjustments when they observe high-redshift objects (for example, they must extend their observations into the infrared region of the spectrum, in order to observe what would be red light in the absence of large redshifts), they can usually measure the amount of reddening in the light from those objects. This amount, in turn, corresponds to an amount of dust along the line of sight, so that astronomers can calculate the extent to which dust has absorbed light of all different wavelengths. With this calculation, they can adjust their observational data to find how the object would look if no interstellar dust affected the passage of its light across billions of light-years.

Thus the differential absorption of different colors of light allows astronomers to recognize the effects of dust grains by spreading the light from a distant supernova into its detailed spectrum. As we have seen, the spectrum reveals the amount by which the Doppler effect has increased all the wavelengths of light. The supernova observers can then "correct" the spectrum for this Doppler effect, using their computers to show how the spectrum would appear if no Doppler shift existed. When they compare this spectrum with the spectrum of a similar supernova much closer to us, they find that the relatively nearby and much more distant supernovae have essentially identical spectra, with the same relative amounts of light at each wavelength throughout the observable spectrum. If dust grains had preferentially absorbed more violet than red light from the distant supernova, this would not be true. The almost perfect match of the (Doppler-corrected) spectra of the nearby and distant supernovae sets a strong upper limit on the density of dust grains in intergalactic space, because astronomers do not detect any additional color-dependent absorption when they look along the much longer line of sight to the far more distant supernova.

COULD GRAY DUST
BE CONFUSING THE COSMOLOGISTS?

However, the limit that astronomers can set on the density of dust grains refers only to the grains that preferentially absorb some colors of light more than others. These comprise the type of dust grains familiar to astronomers. But what if the universe contains significant amounts of a type of dust, still unknown to the experts, that absorbs light of all colors to nearly the same degree? Astronomers call this hypothetical constituent of the cosmos "gray dust," using the word "gray" to denote dust grains that absorb all colors of light equally, so that the dust simply reduces the apparent brightness of a distant object without affecting its color.

Gray dust poses a significant, if hypothetical, problem to astronomers. They can accurately correct each supernova's spectrum in accordance with the observed amounts of interstellar dust along the line of sight so long as they deal with "conventional" dust particles, the amounts of which they can determine from the observed amounts of reddening that the dust grains produce. If the absorbing dust in a far-distant galaxy has the same properties as the dust grains in the Milky Way, it will redden light while it absorbs some of it in just the same way that our local dust does. This reddening allows astronomers to deduce the dust's existence, to calculate its density, and to correct for its effects on starlight. But what if interstellar or intergalactic space contains dust that does not generously reveal itself through the preferential absorption of starlight? What if some dust absorbs light of all colors in equal proportions?

THE WRATH OF AGUIRRE: HOW CAN WE
DETERMINE WHETHER GRAY DUST HAS MADE
DISTANT SUPERNOVAE APPEAR DIMMER?

Although the two competing teams of supernovae experts discussed the possibility of gray dust and presented reasons why this dust probably does not exist in significant amounts, a graduate student at the Harvard-Smithsonian Center for Astrophysics, Anthony Aguirre, has embarked on the most thorough exploration of whether gray dust might explain the supernova observations. If gray-dust particles exist, their properties must differ significantly from those of the dust already detected in interstellar and intergalactic space. Those known dust particles, believed to consist either of pure graphite (carbon) or of silicates (silicon-oxygen compounds), perhaps coated with molecules rich in carbon, preferentially absorb violet light because short-wavelength light has a greater tendency to interact with the molecules in the dust grains. What sort of dust grains, we may ask, would show no wavelength preference in interacting with the different colors of light striking the dust?

Aguirre first hypothesized that these grains could be rapidly rotating, elongated, roughly cylindrical graphite fibers, each about one-tenth of a millimeter long, with a diameter less than one-tenth of the cylinder's long dimension. Graphite grains of this size, considerably larger than the dust already found floating among the stars and galaxies, would interact with light waves of all wavelengths in nearly the same way, qualifying as gray dust by our definition.

A sprinkling of this gray dust throughout the cosmos would act as a nearly transparent screen, absorbing only a tiny percentage of the light passing through every billion light-years of intergalactic space. That percentage would remain constant over all the wavelengths of visible light, allowing the absorption by gray dust to mimic the effects of placing far-distant supernovae at distances somewhat greater than the distances implied by a nonaccelerating universe. If Aguirre's hypothesis were to prove correct, then what has been deemed a cosmological revolution amounts to only the discovery that intergalactic space contains numerous gray-dust particles—an explanation we may call "the wrath of Aguirre" in homage to Werner Herzog's classic film *Aguirre: The Wrath of God*, in which the hero comes to a bloody and catastrophic end. Astronomical issues, however, have never been settled by mortal combat, unless we count the unequal contests in sixteenth-century Europe after which astronomers were burned at the stake for holding opinions offensive to the religious authorities. Gray dust hardly seems worthy of such excessive means of deciding a dispute, even though some careers, as well as the worldwide currents of cosmological thought, may depend on how much gray dust floats throughout the universe. Can we find a way to determine whether gray dust exists in quantities sufficient to have confused astronomers and cosmologists?

To answer this, we first must ask how much dust would be required to have fooled the astronomers in the way that Aguirre suggests, and we then must turn to the issue of whether or not intergalactic space could have acquired sufficient carbon to form this many dust particles. Like all the elements heavier than hydrogen and helium, carbon comes from the nuclear furnaces at the centers of stars. When some of these stars explode as supernovae, they distribute the nuclei they have forged in their nuclear furnaces throughout nearby space. Given what astronomers know, and have extrapolated, about the history of nuclear fusion in stars since the big bang, could 12 billion years of nuclear fusion have produced sufficient carbon to make Aguirre's hypothesis seem reasonable?

Yes, they could—barely. Aguirre calculates, and other astronomers agree, that the hypothetical graphite grains of gray dust would have consumed less than half of the universe's total carbon production since the big bang. This would leave plenty of carbon to explain the amount that astronomers have detected in stars, in cold objects such as planets, and in the interstellar dust grains they have analyzed. A more subtle objection arises from the fact that exploding stars, like all stars, appear almost exclusively within galaxies, so that it is not at all clear that they could ever have blasted much or most of their carbon into the broader realms of intergalactic space to make the gray dust. Astronomers can detect clouds of intergalactic matter when the matter absorbs some of the light from distant galaxies and quasars. By measuring the spectrum of the light from these distant objects, they can determine the abundances of elements heavier than hydrogen and helium. These measurements indicate only relatively tiny amounts of carbon, oxygen, and other elements made in supernovae—not enough to make sufficient gray dust to explain the supernova observations. To this objection, other astronomers add the observation that the processes by which carbon atoms combine into much larger dust particles would not have operated efficiently in the low-density intergalactic environment, partly because of the extremely low density and also because ultraviolet radiation would tend to destroy any dust grains that were made.

Several months after his first speculation about the nature of gray-dust particles, Aguirre pointed out another possibility. Within galaxies, dust grains come in a variety of sizes, all microscopic but some far larger than others. The larger the dust particle, the grayer it will be, capable of absorbing all colors of light in nearly equal amounts. If dust makes its way from galaxies into intergalactic space, it probably does so because hot stars in galaxies emit ultraviolet radiation in sufficient amounts to expel large numbers of the dust particles. This radiation will destroy some of the dust particles by evaporating their surface layers ("sputtering" is the technical term) to the point that the dust disappears completely. This destruction should occur preferentially for the smaller particles, whereas the larger ones will find themselves, perhaps somewhat reduced in size, floating between galaxies. If intergalactic space contains significant numbers of the largest particles and none of the smallest, then once again a gray-dust situation could arise, capable of explaining the supernova results without resort to a cosmological constant.

RESOLVING THE ISSUE OF GRAY DUST

Gray dust or energy hidden in space—which has the real universe chosen? If no one cared about the difference, we could simply await better observations. The question of how the cosmos could have made sufficient gray dust to explain the supernova observations need not be addressed until and unless we obtain more definitive information about the amounts of gray dust in and between the galaxies. If the dust exists, astronomers must find ways to explain how it was produced; if not, they can relax over this particular issue.

Fortunately, astronomers have at least two good ways to attack the issue of gray dust, one already in their grasp and reasonably indicative, the other a few years in the future and capable of yielding definitive results. If gray dust lies among the galaxies, we may reasonably assume that it does not have a perfectly smooth distribution in space. Instead, regions with higher and lower than average densities of dust should exist, as is certainly the case for the interstellar dust that astronomers have detected. An inhomogeneous distribution of dust should broaden the distribution of the peak luminosities observed for the high-redshift supernovae, because, purely by chance, the light from some of them will pass through relatively dense regions containing gray dust, and the light from others will traverse comparatively rarefied regions. As a result, the supernovae whose light has undergone relatively large amounts of absorption by gray dust will appear dimmer than supernovae at about the same distance that reach the same peak luminosity. This difference would appear as an additional spread in the Hubble diagram—a spread greater than that found by the supernova experts. In order to explain the supernova results without a cosmological constant, gray dust must reduce the average brightness of the high-redshift supernovae by 25 percent. In that case, if the distribution of gray dust in space follows a pattern similar to the distribution of ordinary dust, the dispersion around the mean value for the peak luminosities of Type Ia supernovae would be plus or minus 40 percent, about double what the two supernova groups observe. This analysis convinces many astronomers, although Anthony Aguirre believes that the distribution of gray-dust particles could conform to the observations if the dust spreads throughout intergalactic space.

THE PATH TO RESOLVING THE ISSUES THAT CONFUSE ASTRONOMERS

Happily, a far better, more discriminating approach not only can resolve the issue of gray dust versus a cosmological constant, but can also deal with the issue of whether systematic differences exist between nearby and distant supernovae. Gray dust and systematic differences can mimic the effects of a nonzero cosmological constant with high precision only so long as we examine distant supernovae within a relatively constricted range of distances. For now, this well describes the results from the two teams of supernova observers, both of which have concentrated, in order to maximize their flow of data, on supernovae with redshifts between 0.3 and 0.7. These supernovae appear in galaxies with distances from the Milky Way between about 4 and 7 billion light-years. If, however, astronomers observe distant supernovae over a much larger range of distances—one that ranges, for example, from 4 billion up to 10 or 11 billion light-years—then cosmological models allow astronomers to disentangle all other effects from the crucial one: the acceleration produced by a nonzero cosmological constant.

This possibility of separating the two effects exists because gray dust and systematic differences between nearby and distant Type I supernovae both produce effects whose amount must increase as the distance to the supernova increases. If, for example, gray dust has a relatively uniform distribution in space, then astronomers expect to see twice as much absorption of starlight, on the average, if they look twice as far through the universe. As a result, the line on the Hubble diagram describing the real universe should deviate from the model universe with Ω_Λ equal to zero by an amount that steadily increases as we look to supernovae with higher redshifts and greater distances from the Milky Way. Similarly, if systematic differences exist between the maximum luminosities of nearby and distant supernovae, then the effects produced by these differences should increase as we observe supernovae at ever-increasing distances.

In contrast, a cosmological constant also produces a deviation from the line in the Hubble diagram that describes a universe with a zero cosmological constant, but this deviation does not follow the simple additive rule that describes the effects of gray dust. When astronomers succeed in observing supernovae with redshifts and distances much larger than those of the supernovae with redshifts between 0.4 and 0.7, the Hubble diagram for the universe actually reverts toward the original line describing a cosmos with no acceleration produced by a cosmological constant. This reversion occurs because as we look farther out in space, we look further back in time, to eras when the cosmological constant had produced a cumulative effect much smaller than at the present time or at times "only" 4 to 7 billion years ago. We can effectively recapture the Hubble diagram for a universe without a cosmological constant by looking so far back in time that we observe epochs when the cosmological constant had produced negligible results.

Astronomers must therefore not rest on their current supernova assets, proud to have discovered a cosmological constant that makes the universe accelerate and cries out for an explanation of why Ω_Λ has a nonzero value. Instead, they must push their frontiers farther into space, to discover and to analyze supernovae at distances roughly twice as great as those so far investigated. Only then can they eliminate the possibility that gray dust has fooled them and show that the runaway universe deserves general acceptance.

How long will this take? When can we expect the supernova observers to extend their work to redshifts sufficient to discriminate between a universe with a cosmological constant and one in which systematic effects, yet to be measured, have perpetrated a cruel hoax?

The systematic effects that could spoil the revolution might appear as differences between the maximum luminosities of nearby and distant supernovae, or they might appear as differing amounts of gray dust that lie along lines of sight markedly different in their extent. In either case, astronomers have an excellent way to prove or disprove their reality: They must observe Type Ia supernovae with redshifts close to 1.2. Supernovae with these high redshifts exploded at times when the universe had only about one-third its present age, so we look back in time nearly 10 billion years when we observe them. At this epoch, the expansion of the universe proceeded almost as if the cosmological constant were zero, whether or not the constant has the nonzero value suggested by the observations of supernovae with redshifts between 0.4 and 0.7.

To detect, to follow, and to analyze the light from these far-distant supernovae presents a challenging task to the experts, for they must deal with Type Ia's seen at as much as four times the distance, and therefore with significantly less than one-tenth of the apparent brightness, of the already faint high-redshift supernovae that have caused the present furor. Both groups of supernova observers plan to meet this challenge: They have obtained significant amounts of time on some of the world's largest telescopes, and they feel confident that during the next few years, they can find at least a few Type Ia supernovae with redshifts as large as 1.2. A reasonably good statistical analysis, however, requires not a few, but at least 10 or 12 supernovae. "I'll say it will be two years before we find the answer," said Adam Riess at the end of 1999. "We're hoping to find about four of these supernovae per year," says Perlmutter, who therefore agrees, in approximation, with Riess's conclusion concerning the time frame for resolving the issue of systematic effects versus a cosmological constant. Relying on astronomers' hard work and some luck, we may therefore anticipate that the first years of the twenty-first century will show us with finality what Type Ia supernovae have to say about the cosmological constant.

What if we could look still further back, past the era of galaxy formation, back to the time when matter in the universe had barely developed even the tiniest tendency toward clumping? This opportunity exists, and it offers cosmologists the chance to derive the most fundamental of all cosmic sums: the addition of Ω_M and Ω_Λ. It is time to turn our attention to the earliest epoch we can observe, the time when matter and radiation decoupled from each other and allowed matter to form galaxies, stars, planets, and ourselves.

CHAPTER ELEVEN

THE COSMIC
BACKGROUND RADIATION

COSMOLOGY CALLS TO THE HUMAN SPIRIT because it glows with significant questions. Why is the universe expanding? What fate lies in store for the cosmos? What form has most of the matter in the universe assumed, and why? Is the universe finite or infinite, and how can we hope to find out? These are only a few of a long list of cosmological issues that induce pleasure in the contemplation and joy in the answering—if only we have the spirit to do so.

Of the myriad unexpected aspects about the universe, contemplated from our position on a small planet circling a representative star, one of the most startling consists of the fact that we can observe a detailed snapshot of one particular epoch early in the history of the cosmos. That epoch, called the "era of decoupling," occurred a mere 300,000 years after the big bang, when light and other forms of radiation first ceased to interact with matter, and thereafter traveled freely throughout the universe. As a result, only the ongoing expansion has affected the radiation, which continuously passes by us, still rich in information about the state of matter at the era of decoupling. By capturing and analyzing some of this cosmic background radiation, astronomers and cosmologists have developed a raw likeness of the ancient universe into a fine image, which grows ever richer in detail as new instruments observe its characteristics. The radiation not only tells us about the state of the cosmos in the deep past, but also reveals the curvature of space itself, which affects the radiation's travel through billions of light-years of distance.

THE GLORY OF THE DETAILS

The history of astronomical investigation of the cosmic background radiation divides conveniently into three stages. In the first, which lasted from the late 1940s until 1964, the background radiation existed in some theorists' minds, but it had no observational verification. Because so little information existed about the cosmos as a whole, and because the radio-astronomical techniques that would eventually detect the cosmic background were in their adolescence, astronomers tended to overlook the importance of the background radiation, both as proof that the big bang indeed had occurred and as the carrier of information about conditions in the universe during the era of decoupling.

The second stage opened with the detection of the cosmic background radiation in 1964 and lasted for a quarter of a century, until 1990. During this period, astronomers made increasingly accurate measurements of the small fraction of the background radiation that can penetrate the Earth's atmosphere, and they used high-altitude aircraft and balloon- and rocket-borne detectors to make their initial observations of the bulk of the background radiation, which our atmosphere absorbs before it can reach lower altitudes. Astronomers faced a challenging task in sending the sensitive detectors needed to observe the background radiation to high altitudes, where rockets could obtain only a few minutes of data, and lower-altitude balloons only a few hours, before descending once again into the absorbing atmosphere. During the 1980s, some of these flights gave tantalizing indications that the background radiation involves far more energy than conventional theories of the early universe could explain. Although later proven to have arisen from instrumental errors, these results fired the imaginations of cosmologists and made all astronomers eager to see a satellite in orbit for long-term studies of the background radiation.

During the 1970s and 1980s, high-altitude flights revealed the motion of the solar system with respect to the cosmic background. The Doppler effect from this motion leads to slightly higher energies in the radiation arriving from the direction in which we are moving and slightly lower energies for the radiation from the opposite direction. The detection of these Doppler shifts helped to confirm that astronomers had indeed detected the primal glow of the cosmos. It also showed that every "local" system of objects, such as the solar system or the Milky Way galaxy, can have its own motion with respect to the cosmic background radiation, which provides a useful reference frame within which local motions can be measured. We might deduce that the cosmic background therefore supplies what modern cosmology cannot allow: a fixed frame that defines motionless space, against which all motion can be measured. This is false, however. The fixed frame of reference that the background radiation provides can furnish only a local delineation of space, useful within a limited volume but in no way providing a framework of the entire cosmos.

Weep not for the lost innocence of invariant space. The cosmic background radiation presents us with something much more useful and marvelous than a peccant perception of space that just sits. Because the radiation comes to us unhindered from a time when the cosmos was only a few hundred thousand years old, it carries the detailed record of conditions that then prevailed in the universe. In Chapter 4, we saw that the radiation from a distant object with a redshift denoted by z left that object at a time when the universe had a fraction of its present age approximately equal to $1/(1 + z)^{3/2}$. The cosmic background radiation has by far the largest redshift of any source ever detected: Its z equals about 1,300, so its wavelengths have all increased more than a thousandfold above the values they had when the radiation began its journey. This enormous redshift has changed what began as light and infrared radiation into the long-wavelength microwave and radio portions of the electromagnetic spectrum. Calculation of the factor $(1 + 1,300)^{3/2}$ shows that the radiation began its journey—in this case, began to travel through space unhindered by matter—when the universe had only about 1/45,000 of its present age, or about 300,000 years. By allowing for the changes produced by the expansion of the universe, astronomers can display more than their usual archaeological ability of understanding the present by perceiving the past. With the cosmic background radiation, they have access to a relic beyond prize, a full-figured portrait of the universe at a young age. The 1990s opened the modern phase of this cosmo-archaeological investigation, and the next decade should put a dazzling seal on what astronomers have so far achieved.

THE *COBE* SATELLITE OPENS A NEW ERA OF BACKGROUND RADIATION STUDIES

In November 1989, cosmologists received the opportunity of a lifetime, when NASA's *Cosmic Background Explorer* (*COBE*) satellite rode a Delta rocket from the Vandenberg launch facility near Lompoc, California, into a polar orbit around the Earth. Many years in the making, *COBE* carried sensitive detectors to study the full spectrum of the cosmic background radiation, now gloriously opened to yearlong studies by placing the satellite above all the absorption produced by Earth's atmosphere. Under the able leadership of John Mather, the chief project scientist, *COBE* soon made highly accurate measurements of the intensity of the cosmic background radiation at all different frequencies and wavelengths. In January 1990, at the American Astronomical Society's meeting near Washington, D.C., Mather announced to an auditorium, packed with astronomers stunned by the quality of the data and gratified by its implications, that *COBE*'s measurements had confirmed in every detail the predictions of the big-bang model of the universe. Gone forever were the phantasmagoria produced by the rocket-borne measurements of a few years earlier; the big bang stood once more as the fully favored model of the cosmos.

Attention now shifted to *COBE*'s second mission, the far more difficult and time-consuming task of searching for *differences* in the amounts of radiation arriving from different directions. These variations arise from differing densities of matter at the time when the radiation decoupled. Thus the fluctuations in the intensity of the cosmic background radiation have permanently captured the record of slightly greater and lesser densities, the first "seeds" that led to clumps of matter such as galaxies and galaxy clusters. Detecting these differences in intensity faced astronomers with a formidable challenge. They had to allow for two well-known effects that directly interfere with attempts to find subtle deviations from uniformity in the radiation. First, the broad band of the Milky Way—the central plane of our Milky Way galaxy—emits copious amounts of radiation at long wavelengths, most of it from warm dust particles. This radiation far dominates the modest intensity of the cosmic background, swamping it with an overlay that extends all around the sky, permanently concealing nearly a quarter of it from our view, no matter how fine a satellite we may send into orbit. Second, astronomers must allow for the Doppler effect arising from our local motion, which they had already detected and could now study more accurately with the *COBE* satellite. This effect could be measured and analyzed with relative ease as the astronomers searched for fine variations in the intensity of the background radiation.

"Fine variations" here refers to deviations from the average measured in a few dozen parts per million. Small wonder that all previous measurements—with ground-based, balloon-borne, rocket-borne, and even satellite-borne detectors (in a now-forgotten Soviet satellite called *RELICT,* the instruments of which were insufficiently sensitive to achieve success)—had searched for, but failed to find, the departures from uniformity that record the beginnings of clumpiness in the cosmos. *COBE* succeeded, however, with three pairs of radio receivers that recorded the cosmic background radiation at three different wavelengths. At all three wavelengths, the first six months of data revealed the long-sought deviations, and the succeeding three years confirmed the early results. In 1992, George Smoot, the leader of the team that searched for variations in the background radiation, stood before another meeting in Washington, D.C., and announced that *COBE* had found what some reporters dubbed the "holy grail" of cosmology: the inhomogeneities in the cosmic background that show the start of cosmic structure building. After Smoot's statement that "if you're religious, it's like looking at the face of God" was clipped to "seeing the face of God" in the popular press, cosmology ruled the news for nearly a week, until the burning of Los Angeles by rioters displaced the glow from the early universe as the public's chief object of attention.

With *COBE*'s success, astronomers redoubled their efforts to use ground-based and balloon-borne detectors to study the cosmic background radiation, probing for finer details in the departures from smoothness. *COBE* had observed the entire sky, but its telescopes had a wide field of view, almost 10 degrees across, nearly twenty times the angular diameter of the full moon. This made good sense in a full-sky survey instrument, which needed *COBE*'s full four-year lifetime (before its coolant evaporated, leaving its detectors swamped by the radiation that the satellite's own heat produced) to make several all-sky maps to be checked against one another. No ground-based or balloon-based instrument could survey the entire sky so well as COBE, not least because some of the sky would never rise above the horizon. The push was on, however, not to make better all-sky surveys than COBE's, but rather to observe small regions on the sky with a much finer angular resolution than COBE's instruments could attain. This desire, which rose to a nearly overpowering passion as the twentieth century was drawing to a close, stems from much more than astronomers' familiar urges to study the cosmos in finer detail than before. The fine details in the background radiation's divergence from homogeneity can tell cosmologists which models of the cosmos are correct, rejecting others as mere inventions of theorists who have missed the universal boat. In particular, measurements of the cosmic background radiation on a fine angular scale can tell us just what every thinking person wants to know: the sum of Ω_M and Ω_Λ.

THE LAST SCATTERING SURFACE AND THE BEGINNING OF STRUCTURE IN THE UNIVERSE

The bold statement that the background radiation can reveal the sum of Ω_M and Ω_Λ leads naturally to the question, How and why is this possible? (If, instead, you find yourself led to the question, Why me?, you have probably probed as deeply into cosmology as you ought to go.) The background radiation can disclose the total of the contributions that matter and the cosmological constant make to omega for fundamentally the same reason that the curvature of space depends on Ω_M plus Ω_Λ: Each of the two omegas represents an amount of energy; their total determines both the curvature of space and the details that appear in the cosmic background.

What details? In their quest to find Ω_M plus Ω_Λ, astronomers strive to determine the "size of the largest scattering volume" at the time of decoupling. With this phrase, they refer to the fact that before decoupling, radiation and matter interacted intimately and often, as each piece of matter scattered radiation that rebounded from it many times per second. This scattering helped to keep both the matter and the radiation in a homogeneous state, possessed of the same characteristics throughout the universe. But by the time of decoupling, the cosmos had developed irregularities on a tiny scale—precisely the deviations from smoothness that *COBE* measured, and those that cosmologists had been desperate to find. Their desperation arose from calculations showing that unless the cosmos had somehow produced inhomogeneities by the time of decoupling, we have no good way to explain how the universe became so highly structured, only a few billion years later. Cosmologists' fundamental models insisted that cosmic structure formation must have arisen as the continued growth of modest fluctuations already in existence. Years before *COBE* orbited the Earth, cosmologists knew that they needed variations by at least five to ten parts in a million; otherwise, their theories of structure formation, if not their careers, were doomed. *COBE* thus not only put the face of God on the cosmic map, but also stamped the seal of approval on the most fundamental assertion of theories that sought to explain how a featureless froth of particles and radiation evolved into the complex structures we see today.

These theories led to a further conclusion: The fluctuations with the largest deviations from homogeneity should be those whose size equals the maximum distance that light could travel during the time between the big bang and the era of decoupling. As was described in our tour through the inflationary universe in Chapter 5, the finite age of the cosmos means that so long as nothing travels more rapidly than light, regions of space can affect one another only if their separation does not exceed the speed of light multiplied by the age of the universe. Once the inflationary epoch had passed, some 10^{-30} second after the big bang, this rule applied to the entire universe. As a result, at the time of decoupling, 300,000 years or so after the big bang, the largest volume within which all parts of the region could affect one another spanned some 300,000 light-years.

When astronomers observe the cosmic background radiation, they detect radiation arriving from all parts of the universe, the distances of which cause the radiation produced there at the time of decoupling to be reaching us just now. This means that the radiation comes to us from volumes of space that were much larger than 300,000 light-years across at the time of decoupling. We have seen that this fact in turn leads to the horizon problem that helped motivate the inflationary model of the universe. On scales larger than the maximum distance that light could have traveled by the era of decoupling, variations from smoothness do exist; they may well have been created at the end of the inflationary era, when the expansion was still proceeding more rapidly than the speed of light. Cosmological theory predicts, however, that the largest deviations from smoothness should appear in volumes within which internal communication existed at the era of decoupling. In particular, cosmologists expect the largest of all the fluctuations to have arisen within the largest volumes that could maintain internal communication until that time. If astronomers examine the cosmic background radiation on different scales of angular distance and determine the amounts of the deviations from perfect smoothness that appear, they expect to find a peak at a particular angular size. With sufficiently accurate measurements of the background, that peak should stand out like a cosmological beacon, silently describing the angular size of the largest volume that had complete internal causality at the time of decoupling.

All very well, the reader may say, but what means this peak to me? A beautiful cosmic verity, promoted by observations of the background radiation at fine angular scales, consists of the fact that the curvature of space determines the angular size on which we now see these largest "surfaces of last scattering," the greatest regions throughout which particles and radiation interacted up to the time of decoupling. We actually know how large these surfaces were—300,000 light-years across—so the problem reduces to finding how large such a surface appears to us *now*. Happily for cosmologists, the answer depends on the sum of Ω_M and Ω_Λ.

If the universe has a positive curvature, which will be true if Ω_M plus Ω_Λ exceeds 1, then the surface of a balloon once again provides a good model for space. If we imagine ourselves on the balloon's north pole, we can see that as we look in any direction past the equator toward the south pole, space (represented by the balloon's surface and nothing more) tends to converge, with the lines of longitude approaching one another. As a result, the last scattering surface appears smaller to us than it would in a flat universe, where Ω_M and Ω_Λ sum to 1, in which space does not converge.

Conversely, in a negatively curved universe, where Ω_M and Ω_Λ sum to less than 1, space tends to diverge, and the last scattering surface has an angular size greater than it does in a flat universe, which in turn is greater than its angular size if we live in positively curved space.

Therefore, all we need do is to find the angular size of the last scattering surface in our universe, the real one. Astronomers already know that these surfaces should have angular sizes of about half a degree, but they need greater accuracy than this to solve the riddle of Ω_M and Ω_Λ. By analyzing the mass of observational data that describes the cosmic background radiation on different angular scales, astronomers can hope to find the precise angular scale at which the largest deviation from homogeneity appears, furnishing them with a good handle on the curvature of space, which depends directly on the sum of Ω_M and Ω_Λ. To understand how this quest proceeds, we must take a finer look at the cosmic background radiation. This inspection will reward us with another graph, the final one we shall encounter among the great graphical depictions of the cosmos.

SEARCHING FOR THE LARGEST DEVIATION

Ever since *COBE* began to expose the deviations from smoothness in the background radiation, astronomers have displayed their results and their predictions concerning these deviations on a graph that shows the angular scale under observation along the horizontal axis and the amounts of the deviations in the vertical direction (see Figure 11.1). Thus the graph displays, for each scale of angular size, the discrepancy between the observed intensity of the cosmic background and the intensity that would be recorded if the background radiation were perfectly smooth, so that as astronomers study different regions on the sky, they would find no inhomogeneities whatsoever.

For historical reasons, astronomers characterize the intensity of the cosmic background radiation with a temperature, which in fact describes the temperature of the universe at the era of decoupling. Thanks to *COBE*'s initial measurements, we know that this temperature equals 2.7278 kelvins (2.7278 degrees above absolute zero, measured with temperature units equal in size to those on a Cel-

FIGURE 11.1   Fluctuations of the Cosmic Background Radiation on
Different Angular Scales

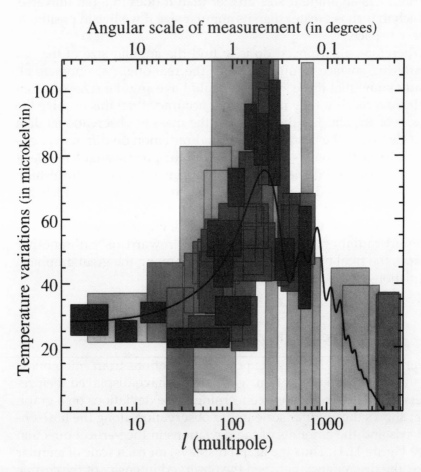

This diagram plots the temperature variations in the cosmic background radiation, measured in millionths of a degree on the Kelvin scale, along the vertical
axis, against the multipole of measurement (lower horizontal axis), which corresponds to an angular scale of resolution (upper horizontal axis). Each experimental attempt to determine the temperature variation appears as a rectangle, within
which the actual values of the variation and the angular scale of observation are
likely to lie. The totality of these measurements of the cosmic background radiation seems to show a peak value for the temperature variations at a multipole
value of approximately 200 to 250. The solid line shows the predictions of a
model universe in which $\Omega_M + \Omega_\Lambda = 1$. (Diagram courtesy of Dr. Wayne Hu,
Institute for Advanced Study.)

sius thermometer but with zero kelvin located at –273.16 degrees Celsius). The differences produced by the Doppler effect, already detected during the 1970s, amount to 3 or 4 millikelvins, that is, to somewhat more than one part in a thousand of the average intensity. On the graph that shows deviations from smoothness at small angles, however, the discrepancies are measured in microkelvins, each one-millionth of a degree on the Kelvin scale. These divergences from the average intensity may appear as slightly greater or slightly lesser amounts of radiation; what counts is the average value of the discrepancy, which runs up the vertical scale almost as far as 100 microkelvins, which would amount to thirty parts in a million of the intensity itself.

The quirky part of the graph that displays the background radiation's deviations from smoothness appears along the horizontal axis, where the angular size of the nonuniformities *decreases* from left to right. To astronomers, this makes good sense, because they are plotting the results from their data analysis, which dissects the observations in terms of "multipoles," a technique first introduced at the end of the eighteenth century by the great French mathematician Joseph-Louis Lagrange. For the radiation that reaches us from a spherical shell, such as the celestial sphere that we imagine to surround us, multipole analysis proceeds by first dividing the sphere into hemispheres, then comparing the amounts of radiation arriving from each hemisphere. This gives the "dipole contribution," the most basic of the multipoles. For the cosmic background radiation, the dipole contribution results almost entirely from the Doppler effect and reveals our local motion with respect to the background.

Multipole analysis not only determines the amount of the dipole (that is, the average deviation from a totally smooth background), but also locates the direction of our motion, by testing all possible ways to divide the celestial sphere into hemispheres and discovering which one produces the largest dipole contribution. The next multipole, called the "quadrupole," divides the shell into four quarters and compares the different amounts of radiation arriving from each of them. In terms of the background radiation's homogeneities, *COBE*'s first significant contribution was its discovery of a quadrupole contribution: the fact that even after we allow for the dipole that our local motion produces, inhomogeneities still remain in the radiation's observed intensity.

After the quadrupole contribution comes the "octupole," measured by dividing the celestial sphere into eighths and comparing the intensity of radiation from each of them. In a similar vein, but with an abandonment of special names, come the contributions from analyzing the radiation by dividing the celestial sphere into sixteenths, thirty-seconds, sixty-fourths, and so on. Cosmologists and mathematicians refer to these contributions as "l-poles," because they use l (the script form of the familiar letter l) to denote each different multipole: $l = 2$ is the dipole, $l = 3$ the quadrupole, $l = 4$ the octupole, and so forth. They plot the deviations from smoothness that they observe as a function of the l-values, each of which measures the deviations on a particular scale of angular size on the sky.

To find the approximate angular size represented by a particular value of l, we may divide 180 degrees by l. Thus the contribution from $l = 10$ refers to angular scales of about 18 degrees, whereas $l = 180$ refers to approximately 1-degree scales, and $l = 1,000$ to deviations on a scale of about 0.2 degrees, or 12 minutes of arc. In Figure 11.1, the l values are plotted along the horizontal axis and the angular size scales on the parallel line that follows the upper boundary of the figure.

Models of how the cosmic background radiation propagates through the universe after the era of decoupling imply that the plot of the relative amount of the deviations at different angular scales should rise to a peak somewhere between $l = 200$ and $l = 500$. The location of this peak depends on the sum of Ω_M and Ω_Λ. More precisely, the peak appears at a value approximately equal to 200 divided by the square root of the sum of Ω_M and Ω_Λ. Thus, if Ω_M plus Ω_Λ equals 1, we expect the peak at about $l = 200$, whereas if Ω_M plus Ω_Λ equals 0.3, the peak should appear at about $l = 200$ divided by the square root of 0.3, close to $l = 400$. Since *COBE* could observe no angular scales finer than about 10 degrees, it could not hope to measure the deviations from smoothness for l-values greater than $l = 18$, and it actually never achieved even this degree of angular resolution. To reach the high l-values necessary to find the value of Ω_M plus Ω_Λ, astronomers need instruments with much better angular resolution than *COBE*'s, significantly finer than 1 degree if they hope to measure the deviations with l-values measuring in the hundreds.

After this exhaustive buildup, the reader cannot help feeling some letdown when confronted with authentic observations of the deviations in the cosmic background radiation at different angular scales. During the past few years, cosmologically minded astronomers have made more than two dozen separate measurements of the background radiation's fluctuations from uniformity at different angular scales. Some of these looked at the long-wavelength component of the cosmic background, which can penetrate the atmosphere and meet the high-resolution gaze of large arrays of radio telescopes. Other observations, looking at shorter wavelengths that experience difficulty in piercing the atmosphere, require more specialized equipment at particular locations, such as the South Pole, where the thin polar air contains far less of the absorbing water vapor that ruins sea-level observations. Still others, made at still-shorter wavelengths, require balloons and rockets to lift detectors above nearly all of the water vapor in the atmosphere. Each of these observations must overcome the interference from all the other sources of radiation in the universe—not to mention our terrestrial noise—that emit photons with the wavelengths that the astronomers are scrutinizing. Each observational result therefore consists not of a single point, as would be true if nothing interfered with the accuracy of measurement, but instead of an "error box" that denotes the region on the graph judged highly likely to contain the facts about the actual universe somewhere inside it. Before leaving this chapter, we shall examine astronomers' hopes for collapsing these error boxes into near-infinitesimal points with a series of new satellites, super-*COBE*s that can study the background radiation with precise angular resolution. For now, however, we must confront the actual data and tease what we can from them.

The graph derived from the totality of experiments consists, as Figure 11.1 demonstrates, of more than three dozen error boxes, through which the line depicting the universe should be drawn. We can look at the horror of this image in at least two ways. On the one hand, only extremely brave scientists will tell you that these results reveal the location of the peak—the multipole with the largest deviation from smoothness—with the precision needed to determine Ω_M plus Ω_Λ. On the other hand, many an astronomer still active in research will tell you that he or she never dreamed, two or three decades ago, that even these boxy results would appear before the next millennium. Both hands wring in delight at the thought that only a few years will bring us results so much better than these that we shall look back to these years and say, I remember when astronomers still had trouble locating the peak angular scale in the fluctuations of the cosmic background radiation.

Even now some truths seem to stand out. The boxes offer strong, though not compelling, testimony that we have detected the peak, because the amounts of the fluctuations drop off fairly convincingly for multipoles beyond 300 or 400, that is, for angular scales smaller than about 0.3 degrees. This fact, if verified, already carries significant weight in cosmological circles. If we believe, for example, that the peak must occur well before the $l = 400$ multipole (i.e., for angular scales significantly larger than those corresponding to $l = 400$), this would rule out models of the universe in which the sum of Ω_M and Ω_Λ falls significantly below unity, say, to values of 0.3 or less. To state the current results in another way, we may say that our best measurements now imply that the sum of Ω_M and Ω_Λ equals a number fairly close to 1, far likelier to equal unity than to fall below 0.4 or to exceed 1.5.

A remarkable complementarity between these observations and those of Type Ia supernovae, which now dominate the cosmological flood of data but may soon cede pride of place to the cosmic background radiation. As we have seen in Figures 7.4 and 9.4, the supernova results provide a long, cigar-shaped region of possibility that rises upward and to the right at a 45-degree angle—and therefore embraces nearly the same value for the difference $\Omega_M - \Omega_\Lambda$ throughout its area. In perpendicular contrast, observations of the background radiation furnish us with another long and cigar-shaped domain, but this one rises upward and to the left, so that $\Omega_M + \Omega_\Lambda$ remains nearly constant inside it. Where the two domains intersect, the true universe lies.

Observations of the cosmic background radiation now locate this region of intersection with values of Ω_M between 0.2 and 0.6 and with values of Ω_Λ between 0.4 and 1.0. If you press an observationally oriented cosmologist to the wall and force him to divulge his best guess at the truth, the numbers $\Omega_M = 0.3$ and $\Omega_\Lambda = 0.7$ will probably emerge from his troubled soul. Considering that just a few years ago, similar pressure to spill his cosmological guts would have yielded $\Omega_M = 0.25$ and $\Omega_\Lambda = 0$, we have seen a sea change in cosmology, an acceptance of the cosmological constant undreamed of for many long decades.

Let us be more precise. The statement above well describes most observationally oriented astronomers, although many of them maintain some reservations about the supernova observations, and still more doubt, as they should, the accuracy of the background radiation observations. On the other hand, almost all cosmological theorists who believed in the inflationary model (and this describes the bulk of them) believed that the total omega equals 1, even when they doubted the existence of a nonzero cosmological constant. This caused them to yearn for, and almost to believe in, some additional dark matter, some hidden component of the universe, some*thing* that would bring the total omega to unity. Some cosmologists modified the inflationary theory so that it did not require that the total omega must equal 1. Other theorists, who once heartily disliked the cosmological constant, came to see it as the joker who can make the party move forward, acceptable even though not socially correct.

THE FUTURE: *MAP*PING THE COSMIC BACKGROUND RADIATION

Nothing separates the nonzero cosmological constant from full acceptance but a couple of years and the proper functioning of NASA's next satellite to observe the cosmic background radiation. If all goes according to plan, this satellite, called *"MAP"* (an acronym for *Microwave Anisotropy Project*, where "anisotropy" denotes deviations from smoothness), will enter an orbit around the sun at the end of the year 2000. Unlike *COBE*, which circled the Earth at an altitude of a few hundred miles, the *MAP* satellite will receive a boost from the moon's gravity to reach an orbit around the sun a million miles outside the Earth's. In one particular location, called the "L2 point" by astronomers, *MAP* can maintain a stable orbit, free from any interference from radio noise on Earth or simply from the Earth's presence, which blocks half the sky for any satellite in a low orbit. From this position, *MAP* will measure the amounts of radiation arriving from different directions to a precision of better than eight parts in a million and with an angular resolution of about one-fifth of a degree, dozens of times better than the angular resolution that *COBE* used.

With this resolution, which corresponds to reaching multipoles close to $l = 500$, *MAP* should be able to look past the first and largest peak in the graph of intensity versus angular size (see Figure 11.2). This peak should occur at a multipole between 200 and

FIGURE 11.2   *MAP* Satellite's Observations of Fluctuations of the
Cosmic Background Radiation on Different Angular Scales

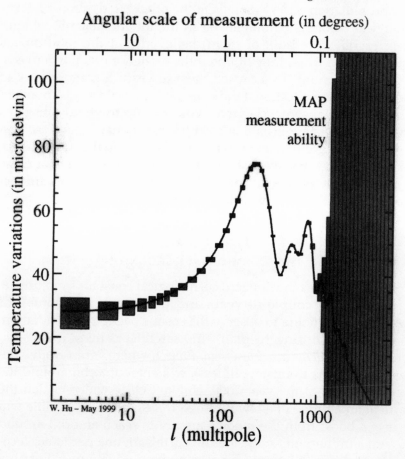

This diagram of (mostly) imaginary results shows the accuracy of observation anticipated for the MAP satellite, to be launched during the Spring of 2001. The two left-most rectangles depict results from the COBE satellite, but the others are purely speculative. The rectangles' sizes show the predicted accuracy of MAP's measurements, which are impressively accurate out to multipole values just past 1,000, corresponding to an angular resolution better than one-tenth of 1 degree. With this accuracy, MAP should be able to establish which model of the universe, with a particular value of $\Omega_M + \Omega_\Lambda$, actually exists. In this diagram, one particular model, for which $\Omega_M + \Omega_\Lambda = 1$, has matched the (imaginary) points perfectly; if past history is a guide, the actual results will not correspond so well to any one model. (Diagram courtesy of Dr. Wayne Hu, Institute for Advanced Study.)

400, depending on the value of Ω, the sum of Ω_M and Ω_Λ. By locating the exact multipole—that is, the angular size scale—at which this peak occurs, *MAP* will allow cosmologists to determine the value of Ω to an accuracy currently unattainable from ground-based and balloon-based observations.

BUT WAIT: THERE'S MORE!

Although the most immediate expectation for the results from *MAP* deal with locating the peak in the graph of fluctuations versus angular size, this satellite and its still more capable successor, the *Planck* satellite, now envisioned for launch in the year 2007, may provide cosmologists with the chance to run the table in their quest to find the parameters that describe the universe. In addition to determining the peak l-value, and hence the angular size of the last scattering surface, an accurate graph will also reveal the actual height of that peak. Just as interesting to cosmologists, it will also show the pattern of dips and subpeaks that should appear on the high l-value or small angular size side of the main peak.

All these dips and peaks are loaded with information about the early universe. The height of the first peak depends on the average density of matter in the universe but not on the size of the cosmological constant. An accurate measurement of this height will therefore provide an independent measurement of Ω_M. The further dips and wiggles reveal facts about what cosmologists call the "spectrum of density fluctuations" at the time of decoupling—that is, the relative amounts by which matter had clumped on the different scales of size that were smaller than the surfaces of last scattering. And these relative amounts depend, so theory tells us, on what the universe was like when it emerged from the inflationary era, 10^{-30} second after the big bang!

Chapter 15 describes in more detail the tasks that await the *MAP* and *Planck* satellites. If all goes well, the first decade of the twenty-first century will markedly increase our knowledge of the details of the cosmic background radiation, advancing cosmologists' abilities to discern how tiny homogeneities became giant clumps of matter. At present, cosmologists are still toiling at this task, not only for its own sake, but also in order to shed light on the total density of dark matter in the universe.

CHAPTER TWELVE

THE BIRTH OF GALAXIES REVEALS THE DENSITY OF MATTER

WHEN GALAXIES WERE BORN

Fourteen billion years ago, give or take a billion or two, the featureless universe began to generate galaxies. Today we see a cosmos full of galaxies, the "island universes" that Harlow Shapley had once pooh-poohed but that contain nearly all the visible matter in the depths of space—and of which Shapley eventually compiled the most complete catalog of his era. Most of a galaxy's visible matter resides in stars, hundreds of billions of them in a giant such as our Milky Way. Other light-emitting material consists of diffuse gas in star-forming regions, lit from within by young stars; of the remnants of supernova explosions; and of the outer layers of old stars, expelled into space and made to shine by ultraviolet radiation from the aged stars' hot cores. Scientific investigation since the early 1970s has shown that our perceptions, honed to the detection of electromagnetic radiation, have misled us by ignoring the bulk of a galaxy's mass: the dark matter, the nature of which remains a mystery and most of which resides in an immense "halo" that surrounds the easily visible portion of the galaxy. Instead of possessing masses roughly equal only to the masses of their stellar content, 100 or 200 billion times the sun's mass, giant galaxies have masses an order of magnitude greater, often exceeding 1 trillion solar masses.

As discussed in Chapter 6, the dark matter in galaxies, even though it provides far more mass than the visible matter, does not furnish the bulk of the total dark-matter content of the universe. The Milky Way, which we may take as a fairly representative galactic giant, has a ratio of dark matter to visible matter that would, if this ratio characterized the entire universe, imply a value of Ω_M equal to about 0.1. For the universe as a whole, the crucial parameter Ω_M equals approximately 0.3.

GRAVITY MADE THE GALAXIES FORM

How do we know this? One path to finding the value of Ω_M, which we shall soon examine, counts the numbers of large galaxy clusters both at the present time and also at times billions of years in the past. Another relies on supernova observations, which yield the value of $\Omega_M - \Omega_\Lambda$. Another relies on observations of gravitational lensing, which furnish the same difference of the crucial parameters Ω_M and Ω_Λ. Yet another calculates how the early universe produced the mixture of light nuclei that we observe today, but this method reveals only the baryonic component of Ω_M, made of matter that participates in nuclear-fusion reactions. Since all these methods involve considerable observational uncertainty, cosmologists remain eager to employ additional means of obtaining Ω_M, in order to reduce the final uncertainty in estimating its value.

By definition, matter produces gravitational forces, the strength of which depends on the mass of the matter and the distance over which the forces exert themselves. The finest way to weigh the totality of matter consists of observing the effects of gravitational forces on other matter. To find the value of Ω_M that describes the entire universe, we need a way to perceive the effects of gravity over distances even larger than those spanned by a galaxy cluster. Big-bang nucleosynthesis, which produced the light elements, serves this purpose for the baryonic component of the cosmos. What method will do the same for the totality of matter?

THE EPOCH OF GALAXY FORMATION
CAN REVEAL THE DENSITY OF MATTER

We can find what we seek in the history of how galaxies formed. If we can reconstruct the past history of the universe, that history, like a "March of Time" newsreel, will reveal details of the crucial epoch when matter pulled itself together to form galaxies and galaxy clusters. Because the gravitational processes that produced galaxies depend on the total density of matter during the galaxy-formation era, the cosmic-history newsreel will allow us to deduce the densities of matter in that epoch. These densities have a direct relationship to the average density today, so a close study of the newsreel will yield the much-sought value of Ω_M.

The difficulty with this approach is that crucial frames are missing from our newsreel. Like the allegedly excised portions of the Zapruder film that captured President Kennedy's assassination in 1963, these missing frames are precisely the ones that might reveal the truth about the cosmic mysteries. We do possess vivid images from the pregalaxy epoch, the era of decoupling. The *COBE* satellite and its successor experiments have provided us with ever more detailed views of the cosmic background radiation as it was about 300,000 years after the big bang. We have even more vivid views of the universe as it appeared a few billion years later, when young galaxies already crowded the cosmos. The light from these young galaxies reaches us with large redshifts, so large that in observing them, we look back more than nine-tenths of the way to the big bang. But we utterly lack any images of the cosmos between the era of decoupling and the time that galaxies began to shine. These missing billion years—quite possibly two or three billion years—lie shrouded in darkness. During that span of time, the background radiation had already decoupled from the matter, so it can tell us nothing of material import; matter itself hid in the night preceding the first stars and galaxies (see photograph on page 184, which reprises the image on page 5).

Cosmologists know that the key to galaxy formation resides in the small variations in density that existed at the time of decoupling, 300,000 years after the big bang. Within the 1990s, improved observations of the cosmic background radiation revealed these fluctuations and allowed astronomers to calculate their sizes. Without these tiny density variations, amounting to only a few dozen

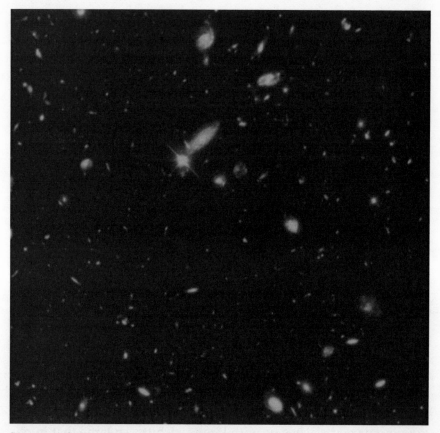

Young galaxies dominate the Hubble Deep Field image. (Photograph courtesy of the Space Telescope Science Institute.)

parts in a million above or below the average, galaxies and stars never would have formed, and we would not be alive to celebrate their contractual success. One of the great frontiers of astronomy, a mystery par excellence that everyone knows can and must be solved, deals with the issue of how extremely modest fluctuations at the era of decoupling evolved into much denser clumps and how these clumps continued to contract until they became galaxies. Everyone knows that the answer to the mystery of galaxy formation lies with gravity, the force expressed by all matter as an attraction for all other matter. No one knows the details, or even the true outlines, of how galaxy formation occurred. Of galaxy formation we see only the fruits, never the efflorescence that produced them. To bridge the gap in time, to restore the missing frames from the cosmic newsreel, we must turn to computer models.

COMPUTER MODELS OF GALAXY FORMATION

In their efforts to deduce what produced the cosmic majesty that we now observe, generations of astronomers have employed generations of improved computers to model the universe's evolution from a near-featureless froth into the extraordinarily complex arrangement we see now, with many different types of galaxies and countless detailed variations on the basic types. This constantly advancing computer game takes models of how galaxies began to form, runs them through computer programs that calculate the gravitational interactions among thousands of separate clumps of matter, and generates maps of the distribution of galaxies in the resulting universe to compare with the actual cosmos. Only the minds of cosmologists and the power of computers can limit the immense variety of possible universes; only carefully gleaned facts about the actual universe can separate the sheep from the goats, the model universes that might yet prove to be real from the models that must be discarded for failing to match the observed distribution of matter in the cosmos.

All of these models begin as the era of decoupling ends, and all use the most recent observations concerning the fluctuations in density that existed 300,000 years after the big bang. In addition to these fluctuations, which astronomers are measuring with ever-increasing precision, the most important parameter that describes each model is the total density of matter at the time of decoupling. The matter providing this total density consists of the components that we met in Chapter 6: hot dark matter and cold dark matter, which may be further deconstructed into various subcomponents, some real and some hypothetical.

The distinction between hot dark matter and cold dark matter proves crucial in the story of how objects formed. These terms distinguish hot dark matter, the constituent particles of which had speeds close to the speed of light at the end of the era of decoupling, from cold dark matter, with speeds much less than the speed of light. In a hot-dark-matter universe, the largest structures should have formed first. The rapid motions of the hot-dark-matter particles would have prevented small clumps of matter from contracting; instead, the particles in these clumps would have been smeared again into a featureless, homogeneous mixture. Larger concentrations of matter would have been less affected by this smearing-out tendency, with the largest and most massive affected the least. In a cosmos of cold dark matter, the formation process would have proceeded in the opposite sense, with small clumps, eventually to become galaxies, the first to form. Galaxies would then have gathered themselves into galaxy clusters, which in turn would have formed superclusters as the universe grew older.

All observational signs point toward a predominance of cold dark matter over hot dark matter. Superclusters appear to be forming only recently in cosmic time: When astronomers look billions of light-years away, into epochs when the universe was less than 10 billion years old, they observe far fewer superclusters within a standard volume of space than appear at distances less than 1 billion light-years. Galaxies, on the other hand, clearly formed in great numbers before the universe was more than 3 or 4 billion years old. The peak of galaxy formation occurred in regions of space with observed redshifts between 2 and 4, which correspond to times when the universe had an age between one-fifth (for a redshift of 2) and one-eleventh (for a redshift of 4) of its present age—in other words, the era between about 1.3 and 3 billion years after the big bang.

Assuming, then, that the cosmos contains far more cold than hot dark matter, the most basic question for the future of the universe remains not, What's the matter?, but rather, How much does it matter? What is the value of Ω_M that gives the best fit between a model based on the inflationary scenario plus hot dark matter and the universe that lies around us? (We must always recall that although the average density of matter has continuously decreased as the universe expands, if we specify the value of Ω_M at the present time, we name the entire history of the average density, because we can easily calculate how the density has changed with time.) The response to this question amounts to a bit of a letdown: The models have not yet reached the point where they can provide anything close to a definitive answer. Too many unknowns still lurk in the formation process, including such fundamental issues as how accurately the galaxies and galaxy clusters trace the actual distribution of matter. Could it be that these luminous objects form preferentially in the regions of highest density, as seems likely from a knowledge of how gravity pulls objects together? If so, how much denser than average should we expect the dark matter to be in and around a galaxy cluster, in comparison with the vast spaces between the clusters? Until questions such as these receive fairly detailed answers, cosmologists who model the universe will have difficulty in naming the single value of Ω_M that best describes the universe on the basis of how galaxies formed. About the best that the modelers can say now is that Ω_M appears to lie somewhere near 0.3, almost certainly exceeds 0.1, and probably falls short of 0.5.

We may also notice that the relative proportions in the model of hot and cold dark matter have an important effect on the outcome. The most successful recipes for galaxy formation contain not only cold dark matter, but also an admixture of hot dark matter, amounting to about one-fifth of the total density, to leaven the mixture. These "mixed-dark-matter" models understandably achieve greater success than pure hot-dark-matter or cold-dark-matter models can, since they provide an additional parameter to be adjusted by the modelers as they seek the best fit with observations. Even so, fulfilling the dream of generating a model cosmos that can exactly match the real universe's distribution of matter seems to lie a few years in the future.

Before we abandon hope of using the formation of galaxies and galaxy clusters to show us the exact value of Ω_M, however, we must pause to examine a related issue that better completes this task. Instead of attempting to compare the entire universe with computer models, we can concentrate on a single class of objects, the large clusters of galaxies, to reach an important conclusion about the average density of matter in the universe.

GALAXY CLUSTERS:
OBSERVATIONS AND FORMATION THEORY

Large clusters of galaxies, of which the Virgo Cluster provides us with the closest and best-studied example, each contain a few thousand member galaxies, spread out over a diameter of about 10 million light-years. In any reasonable model of how structure formed, we naturally expect that more clusters exist now within a standard volume of space than was the case many billion years ago. Therefore, a comparison of the numbers of galaxy clusters at relatively nearby distances—out to a billion light-years or so— with the numbers seen at distances of many billion light-years should reveal larger numbers of clusters in nearby regions, once we compare equal volumes of space.

And, indeed, just this effect appears in the data. Furthermore, the relative numbers of galaxy clusters existing in the present epoch and in eras billions of years ago provide a relatively sensitive means of estimating the average density of matter. In a high-density universe, one with Ω_M close to 1, the formation of clusters should have been proceeding ever more rapidly. As a result, within a standard volume of space, astronomers would expect to find far more clusters with low redshifts than they would when they look for clusters with z close to 1. By counting the clusters with redshifts close to 1, astronomers are determining how many clusters had already formed when the universe had about 40 percent of its present age.

The results now seem clear: Astronomers do not see a great difference between the numbers of clusters with low redshifts and those with redshifts close to 1. The observations effectively eliminate the possibility that Ω_M could have a value close to 1, and they single out values close to 0.3 as the most likely ones for Ω_M. With values of Ω_M much lower than this, the cosmos could not have formed the number of galaxy clusters that we see at what effectively amounts to the present time; with values much higher than 0.3, we would expect to see almost no clusters whose redshifts take us to eras when the universe was five to eight billion years old.

Note that although we might think that observations of galaxy clusters can reveal only the density of matter *within* the clusters, this conclusion would be mistaken, because we are concentrating on the processes that formed the clusters. These processes involve not only the material that became part of a galaxy cluster, but also the matter that did not. The density of matter outside the cluster played a crucial role in determining how many large clusters of galaxies formed at different epochs in cosmic history. The value of Ω_M cited in the previous paragraph therefore refers to the universe as a whole. During the next few years, better computer models, together with the new survey of galaxies described in Chapter 15, will produce improved estimates of Ω_M, based on the big picture of galaxy formation. For now, however, the most accurate way to estimate the value of Ω_M relies on the observed numbers of galaxy clusters, coupled with our understanding of how these clusters formed.

WHAT CAUSED THE TINY FLUCTUATIONS THAT PRODUCED US?

The realization that all the structure in the cosmos has grown from tiny fluctuations in the density of matter from place to place leads naturally to the question, What produced these minuscule variations? To answer this question amounts, in a cosmological sense, to explaining how the universe grew from no structure into complexity.

If we accept, as nearly all cosmologists do, the inflationary model of the universe, then we must seek the modest seeds of galaxy formation in the inflationary epoch, the period soon after the big bang when the cosmos expanded so many times that it flattened itself completely, producing the $\Omega_M + \Omega_\Lambda = 1$ universe that we see today. As discussed in Chapter 5, the inflationary era also produced a smooth universe, because a causally connected region of space expanded immensely during the inflationary era, preserving its smoothness as it did so.

Total smoothness never existed and never will. Within even the smoothest-seeming gas, a froth of individual particles bounces endlessly from place to place. Quantum theory implies that the density of these particles cannot be exactly the same everywhere or at all times. Indeed, quantum theory predicts that density fluctuations must occur in any gas. Even though these density variations begin with incredibly small sizes, inflation does such an excellent job of enlarging everything in the cosmos that the distances over which the fluctuations emerge from the inflationary era grow to considerable sizes.

Notice how smoothly the term "quantum theory" offered itself as the explanation for why the universe ever formed clumps of matter. As we shall see in Chapter 14, this sleight of hand pales in comparison to using quantum theory to explain the origin of the universe itself. At present, we may reasonably accept the verdict of physicists who have tested and retested quantum theory, verifying that its mathematics well explains how particles interact on the smallest scale of sizes. We can certainly accept the notion that at these size scales, matter should *not* behave in accordance with our intuition and experience, which refer to much larger sizes. Among other things, experience tells us that all particles can be sliced into smaller ones. Quantum theory rejects this rule, since it implies that the sliced and diced particles cannot be the elementary ones, incapable of further division. A theory of elementary particles must differ in fundamental ways from the ones that we use to describe giant assemblages of those particles. Once we accept this principle, all that remains is to marvel at just how radically the quantum theory's description of reality differs from what we expect. Among those differences, and central to the theory, we find the assertion that nothing can remain absolutely at rest. Small wonder, then, that quantum theory predicts that small fluctuations in the density of matter will inevitably emerge from the era of inflation.

From quantum theory we may turn to the other great innovation of twentieth-century physics: Einstein's general theory of relativity. From this theory comes another possibility for determining the average density of matter in the universe, based on the well-established fact that gravity bends space—a phenomenon as odd as anything the quantum theory tells us.

CHAPTER THIRTEEN

GRAVITATIONAL LENSES BEND THE COSMOS

THE RACE TO DISCOVER THE SECRETS of the universe now concentrates on attempts to determine the contributions made by all types of matter and the cosmological constant to the total energy density of the universe. The symbols Ω_M and Ω_Λ represent these contributions, measured as fractions of the critical density of matter defined in Chapter 4.[1] Among the techniques that astronomers use to determine these two crucial parameters, we have examined measurements of the deviations from smoothness in the cosmic background radiation, which effectively measure the sum of Ω_M and Ω_Λ, and supernova observations, which determine the difference between Ω_M and Ω_Λ. If we knew both this sum and this difference, our quest would be over, since we could immediately find $2\Omega_M$, and thus Ω_M itself, by adding the sum and difference of Ω_M and Ω_Λ, and we could then find $2\Omega_\Lambda$ by subtracting the difference of the two parameters from their sum. We have also seen that attempts to model the evolution of galaxies and galaxy clusters yield a direct, though not very accurate, estimate of Ω_M, placing it in the vicinity of 0.3.

Eventually, astronomers who measure the cosmic background radiation and make observations of supernovae will obtain suffi-

---

[1]Using Einstein's famous equation, energy equals mass times the speed of light squared, we can convert a density of energy into a density of matter by dividing the amount of energy per cubic centimeter by the square of the speed of light.

ciently accurate data to satisfy even skeptical theorists that they must discard some models of the universe as invalid. Only a limited subset of those models, determined by specific values of Ω_M and Ω_Λ, will conform to the observational results. For now, this happy state remains a vision to be achieved in the near future. Before then, astronomers must rely on other weapons in their arsenal to resolve the conflict of the cosmological parameters.

GRAVITATIONAL LENSING

Among the methods by which astronomers can hope to determine the values of Ω_M and Ω_Λ, one of the most amazing relies on gravitational lensing, the focusing of starlight by gravitational forces. Gravitational lensing occurs when an object with mass—a large galaxy, for example—bends light passing by it, a phenomenon first conceived in the mind of Albert Einstein several years before astronomers observed it in the sky. Einstein predicted that gravitational forces will bend light rays and also the amount of this bending. When astronomers verified that this bending occurs, Einstein became famous worldwide—a scientist who had predicted the curvature of space before observations revealed it.

Two complementary views of the effects of gravity can explain why gravitational forces bend light rays. The view that Einstein preferred states that space itself bends and distorts as the result of gravitational forces. More precisely, gravitational forces reveal themselves by distorting space. In Einstein's view, planets orbit the sun because the sun's gravity bends space, making the planets roll through bent space like marbles rolling around a funnel-like depression on a table. Light rays generally follow straight paths, but if these paths happen to pass through bent space, they must deviate from straight-line trajectories because no such trajectories exist in the bent region. The light rays will bend more if they approach closer to an object with mass or if they pass by a more massive object.

The complementary approach to Einstein's regards space as flat, as we intuitively believe it ought to be, and sees gravitational forces as acting upon light, even though light consists of massless particles. In either approach, any object with mass affects not only other objects with mass, but also light rays, which have no mass at all but nevertheless deviate from the straight-line trajectories they

FIGURE 13.1 Gravitational Deflection of Light by the Sun

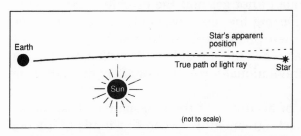
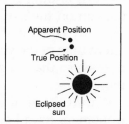

(not to scale)

Einstein's general theory of relativity predicts that the sun's gravity will bend space, so that starlight passing close by the sun will not travel in a perfectly straight line, but will instead bend toward the sun. The gravitational deflection of starlight will make the star seem to lie further from the edge of the sun than is the case. During the total solar eclipse of May 29, 1919, Einstein's prediction received triumphant vindication. (Diagram courtesy of Jon Lomberg.)

would have if no objects with mass acted on them. Our eyes and brains have no aptitude for bent space. When we use them to locate an object, we perform the perfectly reasonable extrapolation of assuming that the object lies exactly in the direction from which its light shines upon us. Einstein showed, however, that this extrapolation will lead us into error in the astronomical arena. As light from a distant source passes by a massive object, its gravitational forces bend the light toward it. The light then appears at a position on the sky shifted outward from the massive object, and we never stop to ask whether the direction from which the light arrives represents the actual position of the object.

Suppose that the sun provides the massive object. Then, Einstein showed, if a distant star lies in just the direction that would cause its light to graze the edge of the sun, the gravitational bending of light will displace the star's image outward by approximately 1.75 seconds of arc. Small though this may be, it corresponds to an angular distance that astronomers can measure—if only they can arrange for the sun to go dark.

In Einstein's era, total solar eclipses provided astronomers with their only opportunities to make accurate measurements of starlight passing close by the sun. On May 29, 1919, as the moon's shadow traced a narrow path of totality across the Earth's surface, British astronomers who had traveled to the islands of Príncipe and Fernando Póo (now known as Bioko) in the Atlantic Ocean used their telescopes to photograph the eclipsed sun together with its surrounding corona of hot gas and the brighter stars close by the sun on the sky. During the next few months, these observers carefully compared their photographs with others taken at a different time of the year, when the sun did not lie in that direction on the sky, so that its gravitational force did not affect the light from those stars.

Within the limits of accuracy of their measurements, the astronomers found exact agreement between Einstein's prediction and their observational results. The stars' images photographed during the eclipse showed displacements outward from the sun by amounts that decreased in proportion to their angular distances from the sun's center, just as Einstein's general theory of relativity said should occur. In Berlin, Einstein awoke to find himself famous. Although his special theory of relativity had made him a household name among physicists' families for more than a decade, the public had largely ignored the theory's implications. Now the times called for a celebration of science and the international sharing of knowledge, symbolized by the British confirmation of a German scientist's theory less than a year after World War I had ended. For the rest of his life, Einstein remained the epitome of a scientific genius, the only scientist many people could name—Carl Sagan and Stephen Hawking rolled into one.

During the six decades following the first observations of the bending of light by gravity, the world of physics and astrophysics saw the development of quantum mechanics, the mass emigration of European scholars, the advent of artificial nuclear fission and fusion, and the discovery of quasars, pulsars, and black holes. Amid these and a host of other events, the "gravitational deflection of light," as physicists came to call it, remained a key proof of Einstein's general theory of relativity but little more, a sort of parlor trick to recount for the public with no practical application to other realms of science. Astrophysicists made increasingly accurate measurements of the sun's gravitational deflection, including its deflection of radio waves; these improved observations likewise conformed to Einstein's predictions and significantly disproved some suggested variations on Einstein's theory of gravity. But in the larger context of the vast cosmos around us, the gravitational deflection of light had little impact—until astronomers began to observe this effect in the depths of space.

THE GRAVITATIONAL DEFLECTION OF
STARLIGHT BY GALAXIES

A giant galaxy of stars, which contains approximately one trillion times the mass of the sun, has an impressive potential for bending light by gravity. If we hope to observe this effect, we must find situations in which a still more distant source of light—another galaxy, for instance, or a quasar, which probably forms the particularly luminous center of a young galaxy—happens to lie almost directly behind the galaxy, as seen from our position in the Milky Way. Since the 1980s, as astronomers have extended their high-resolution view of the cosmos to successively greater distances, they have found, simply as the result of chance applied to a larger number of cases, an ever-growing number of suitable situations. These lineups allow the closer object to deflect light from the more distant source in a variety of possible ways, all of which possess interest for astronomy.

If the distant source appears as a point, as is true for quasars, rather than as an extended object, then the intervening massive object (almost certain to be a galaxy) can produce two, three, or more separate images of the source as it deflects the light around its different sides. If the distant source happens to lie exactly behind the object, and if the object has spherical symmetry, these multiple images will merge into a single ring of light, called an "Einstein ring" in honor of the man who first perceived the possibilities of the gravitational deflection of light. If the centering becomes imperfect, or if the lensing object is asymmetric, we may see part of a ring, called an "Einstein arc" or a "gravitational lens arc." All of these possibilities have now been observed as reality.

In the cases in which our view of the situation does not have a sufficiently crisp resolution to reveal separate sources, an Einstein ring, or a gravitational lens arc, astronomers can still detect the effects of the gravitational deflection of light. The gravitational bending of light often focuses light rays, making the source appear brighter than it would if no intervening object existed to exert its gravitational effects. Astronomers use the term "microlensing" to describe situations in which they observe the brightening of an object caused by gravitational lensing while observing only a point of light and not any of the features, such as an Einstein ring or a lens arc, that the lensing produces.

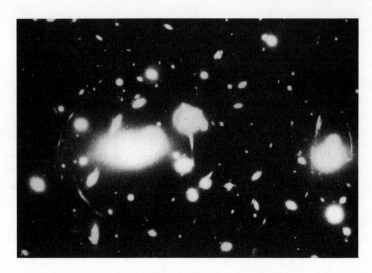

The bending of light by gravitational forces appears on a cosmic scale in this photograph of a cluster of galaxies taken by the Hubble Space Telescope. The largest and most massive galaxies in the cluster bend light from still more distant galaxies, producing a complex set of Einstein arcs, most noticeable near the left-hand edge of the image. (Photograph courtesy of the Space Telescope Science Institute.)

USING GRAVITATIONAL LENSING
TO DETERMINE THE COSMIC PARAMETERS

In recent decades, gravitational lensing by galaxies has made the transition from a rare phenomenon, more to be admired than employed, to a well-understood method of gathering cosmological information. In several cases, for example, astronomers observe two quasars close together on the sky, with essentially identical redshifts in their spectra. Closer inspection reveals that an intervening galaxy has made a single quasar appear to us as double. In some cases, astronomers can see the galaxy that lies almost directly between ourselves and the quasar; in others, they detect only the effects on the quasar's light produced by the intervening galaxy's bending of the space through which the light passes.

The amount of lensing that occurs within a standard volume of space depends upon the number of objects with masses comparable to galaxies' masses, but it does not depend on whether or not astronomers can see the objects that produce the lensing. Indeed, gravitational lensing detects all objects with masses comparable to galaxies, whether or not they are galaxies. The fact that gravitational lensing finds mass without regard to the form that the mass may take makes this technique a highly useful way to measure the total density of matter in the universe. In the bygone days when the cosmological constant seemed to be zero, astronomers expected that gravitational lensing might soon provide an elegant and direct means of determining the density of matter.

Once we allow the possibility of a nonzero cosmological constant, the situation becomes a bit trickier. Like the interpretation of the observations of high-redshift supernovae, gravitational-lensing results implicate both the total density and the cosmological constant. In the case of supernovae, astronomers observe how these two factors, expressed in terms of Ω_M and Ω_Λ, affect objects' distances and thus their apparent brightnesses. A larger total density of matter lessens these distances by allowing less expansion than a smaller density would, whereas a nonzero cosmological constant increases the distances by tending to accelerate the expansion.

The mathematical details of the interplay between Ω_M and Ω_Λ in gravitational lensing do not obey quite so simple a rule as occurs for the case of supernovae brightnesses. As a result, astronomers require more gravitational-lensing situations to reach a particular accuracy with this method than they do high-redshift supernovae to reach the same level of accuracy in determining the cosmic parameters. At this time, observations of supernovae have provided more exact values for the crucial cosmic parameters than gravitational lensing can. The best current results from observations of gravitational lensing show that if we assume that the sum of Ω_M and Ω_Λ equals 1, then the value of Ω_Λ almost certainly lies below 0.75, probably falls below 0.65, and likely lies below 0.5. (The qualifying adverbs have precise statistical meaning, but we may skip past these details as we await future analyses of the data.)

These early results have some overlap with the conclusions based on observations of high-redshift supernovae, but they also hint at a possible discrepancy between the two methods. The supernova results imply that Ω_M minus Ω_Λ equals approximately -0.4, so that if Ω_M equals about 0.3, as other observations suggest, then Ω_Λ has a value close to 0.7. Both approaches to determining

Ω_Λ easily allow for a value of 0.65 or 0.6, but if Ω_Λ turns out to be significantly less than 0.5, something must be wrong with the current supernova results, which imply that Ω_Λ should not be less than 0.5. If Ω_Λ does have a value that low, some error must exist in either the supernova or the gravitational-lensing data, or in both. Indeed, if astronomers discover that Ω_Λ has a value well below 0.5, cosmology will experience a great swing back toward the old days, when Ω_Λ had a value of 0, simply because once a significant portion of the current value of 0.7 estimated for Ω_Λ has been removed, a sort of unscientific momentum will suggest that the entire cosmological constant may well disappear upon closer scrutiny.

Fortunately for the implications of gravitational lensing, significant improvements in this technique should appear within a relatively short time. After working for years with observations of the gravitational lensing of visible light from far-distant quasars, the gravitational-lens experts have shifted their focus to analyzing lensing events observed with radio waves. The bending of space naturally affects the radio waves emitted by distant, powerful radio sources, the quasars and radio galaxies, just as it does light waves. Dust grains in interstellar and intergalactic space, which were described in Chapter 10 as the bane of visible-light observations, absorb and redden the light from quasars and galaxies, confusing astronomers in their attempts to count the number of gravitational-lensing events. Since dust has essentially no effect on radio waves, this portion of the spectrum offers the current best hope to find the cosmological parameters by observing gravitational lensing.

We have now exhausted cosmology's basic methods for determining Ω_M and Ω_Λ, and perhaps the reader as well. The time has come to ask what can possibly explain the existence of a nonzero cosmological constant, if such turns out to be the case for the universe in which we live. The answer may lie in a sweet approach to the big picture, which cosmologists call the "anthropic principle."

Summary of Observational Results for the Values of Ω_M and Ω_Λ

Method	Result (statistically most likely)
Supernova observations	$\Omega_M - \Omega_\Lambda = -0.4$
Cosmic background radiation	$\Omega_M + \Omega_\Lambda =$ approximately 1
Formation of structure	$\Omega_M = 0.3$
Numbers of galaxy clusters	$\Omega_M = 0.2$
Gravitational lensing	$\Omega_\Lambda =$ about 0.5 if $\Omega_M + \Omega_\Lambda = 1$

Within the limits of observational inaccuracies, all these results fit with the values $\Omega_M = 0.3$, $\Omega_\Lambda = 0.7$

CHAPTER FOURTEEN

HOW CAN WE EXPLAIN THE COSMOLOGICAL CONSTANT?

IF THE COSMOLOGICAL REVOLUTION OF 1998 survives close scientific scrutiny, we must somehow reconcile ourselves to the existence of a universal cosmological constant, an energy that lurks in every cubic centimeter of empty space and causes the cosmos to expand ever more rapidly as time goes by. To explain this fact requires more than the usual suspension of disbelief that allows us to accept concepts such as the origin of the universe—all of space and the matter within it—in a sudden big bang, or the expansion of the cosmos without needing anything to expand into.

At the professional level, the existence of a nonzero cosmological constant raises two enormously difficult issues. This excites the cosmology pros, who are bored by the easy ones. Responding to one of these questions, cosmologists and particle physicists have striven for decades to explain why the cosmological constant might not be zero and, if so, why the constant should have a particular nonzero value. Particle-physics theories appear to offer at least a partial explanation of a nonzero value, but the values that these theories predict lie so far from reality that an outsider might reasonably judge them to provide no explanation at all. The finest theories of how elementary particles interact imply that the cosmological constant could either equal zero or have a value at least

10^{120} times greater than the value implied by the recent supernova observations. To err by a factor of 10^{120} is to be gloriously wrong. Cosmological theorists must strive more vigorously for an explanation based on physics only dimly glimpsed at best, and they are doing so.

Why are Ω_M and Ω_Λ So Close in Value?

Whatever explanation cosmologists eventually produce must deal with the second great issue invoked by a nonzero cosmological constant. As we have seen, a cosmological constant makes the universe expand ever more rapidly. Working in the opposite sense, the mutual gravitational attraction of all objects with matter opposes the expansion. The effectiveness of this opposition depends on the average density of matter, which steadily declines as the universe expands. In contrast, the cosmological constant generates an unvarying tendency to increase the rate of expansion.

During the epochs soon after the big bang, the universe was rapidly expanding as the result of the swift kick, as we may imagine it, that all of space had received at the moment of the big bang. Then as now, two impulses—gravitation and the cosmological constant—influenced the expansion in opposite ways. In the early universe, matter had a density far greater than the present value. As a result, the gravitational effect on the expansion completely dominated any contribution from the cosmological constant. For its first few billion years, the expansion evolved in almost exactly the same way—slowing a bit more each year as gravity did its best to pull the cosmos together—as it would in a universe with no cosmological constant.

Through billions of years, the density of matter declined and, with it, the ability of gravitation to slow the expansion. In a universe with Ω_Λ equal to 0, gravitation would still be the only game in the cosmos, and the burning question would remain whether the ever-weaker influence of gravity might nevertheless eventually reduce the expansion rate to 0. As we have seen, that question correlates to whether Ω_M exceeds 1. If the cosmological constant does not equal 0, its importance in comparison with gravity's effects will steadily grow. In algebraic terms, in a cosmos where Ω_M and Ω_Λ sum to 1, Ω_M will fall and Ω_Λ will rise continuously, in a manner that keeps their total equal to 1. Recall that Ω_M and Ω_Λ not only effectively measure the contributions that matter and the cosmological constant make to the curvature of space, but also describe their effects on the expansion of the universe. In a flat universe, one in which Ω_M plus Ω_Λ equals 1, the first one or two billion years after the big bang saw Ω_Λ less than 0.01 and Ω_M greater than 0.99. This gives mathematical expression to the statement that the cosmos then behaved as if the cosmological constant were 0.

The passage of time has tilted and finally flipped the relative importance of Ω_M and Ω_Λ. According to what the supernova data tell us, for a few billion years after the big bang, Ω_M remained significantly larger than Ω_Λ. About 10 billion years A.B.B. (as we may abbreviate dates in cosmic history), Ω_Λ came to equal Ω_M. About four billion years later, at the present time, the cosmological constant has taken the lead: The supernova observations imply that Ω_M equals 0.3 and Ω_Λ equals 0.7. If the observations are correct, then 10 billion years from now, Ω_M will have fallen to about 0.1, and Ω_Λ will have risen to 0.9. From then on, cosmologists can almost neglect Ω_M completely in calculating the future of the universe: Ω_Λ will rise ever closer to 1, and the expansion will proceed ever more rapidly, driving the galaxies, which will consist ever more completely of burnt-out stars and dark matter, to ever-greater separations.

We therefore live at a special moment in the history of the universe, producing the so-called Kerrigan problem, the brief span of cosmic time when Ω_M and Ω_Λ have roughly comparable values. Even though this interval lasts for some 20 billion years, from 5 billion years after the big bang to a future epoch some 25 billion years A.B.B., we may rightly call this brief in comparison with the far future of the cosmos, which extends into a dull and empty infinity. To scientists, the natural way to view these extents of time employs a logarithmic scale, which proceeds by factors of ten. On such a scale, the first billion years after the big bang occupy the same portion of the time line as the first 10 billion years A.B.B., which in turn receive as much room as the first 100 billion years A.B.B. In this perspective, the span of 20 billion years during which Ω_M roughly equals Ω_Λ includes just a bit more than one factor of ten—a single time interval along a time line with an infinite number of intervals.

How, then, can we explain the most upsetting implication of the supernova results, the notion that we live at a special time in the history of the universe, the era when the cosmological constant and the density of matter make roughly equal contributions to the curvature of space? One explanation sees this outcome as the result of chance: We must occupy some position along the line of history, and we just happen to find ourselves at an interesting location. To some cosmologists, this simple explanation suffices completely. Many experts, however, feel a deep upset from the growing realization that Ω_M roughly equal to Ω_Λ may well describe our moment in cosmic history (give or take 10 billion years) and no other. For some of them, the most reasonable explanation of the fact that we live in interesting times lies in a concept that seems to elevate the significance of our own existence on Earth: the anthropic principle.

The Anthropic Principle: Fact or Folly?

Like so many theoretical constructs, the anthropic principle comes in at least two varieties, first named and described by the cosmologist Brandon Carter. The first of these, technically called the "weak anthropic principle," states that because living organisms require certain physical conditions—for example, densities and temperatures that allow large molecules to form and to interact—the existence of any living organisms must violate the Copernican principle to some extent. We could not expect to observe the universe as it was during the first billion years after the big bang, when galaxies and stars had barely begun to form, simply because life could not have evolved anywhere to produce beings capable of cosmic observations. Thus the weak anthropic principle, which we shall abridge as simply the anthropic principle, explains why we do not find ourselves in any of the time intervals less than one billion years A.B.B., of which an infinite number exist on our logarithmic scale.

Similarly, the anthropic principle explains why we do not find ourselves alive at a time greater than, say, one trillion years A.B.B. At that time, almost all the stars in the universe with masses comparable to the sun's will have burnt themselves out, lowering the curtain on any life that may exist on planets orbiting these stars. To a rough approximation, subject to revision as we learn more about the processes that generate intelligent life in the universe, the anthropic principle singles out the span of time between approximately one billion and one trillion years A.B.B. as the interval in which any intelligent forms of life could exist. By what now seems only coincidence, this time interval includes the span of history during which Ω_M and Ω_Λ had anything close to comparable values: At earlier times, Ω_M was more than 100 times greater than Ω_Λ, and at times after one trillion years A.B.B., Ω_Λ will be more than 1,000 times greater than Ω_M.

Does the anthropic principle assert more than that we're here now because we happen to be here now? Yes, it does. The principle ties the physical conditions in the cosmos to the physical requirements for life to exist, and it locates the latter within a particular span of time in cosmic history. Thus our own existence—more accurately, the existence of any form of life capable of contemplating these issues—locates the present moment within a wide but specific range of time after the big bang. For reasons that may be coincidental, or may prove to embrace some of the greatest discoveries in physics of the next century, that range of time includes the eras when the cosmological constant and gravitation vie on roughly equal terms to dominate the cosmic expansion. Let us therefore pause for a paean to life. Despite its name (from *anthros,* the Greek word for "man"), the anthropic principle actually amounts to the intelligent-life principle: We find ourselves alive in a cosmos with roughly equal values of Ω_M and Ω_Λ because those values occur at a time when intelligent life can flourish.

Those who favor the anthropic principle may love or despise the "strong anthropic principle," the notion that the universe must obey the physical laws that we observe because only those laws, or others close to them, allow life to exist. If, for example, the cosmological constant had a much greater value than that implied by the supernova observations, the runaway universe would accelerate so much more rapidly that long before life could arise, everything in the cosmos would find itself at such immense distances from everything else that, for example, no star or planet could ever have formed. (This, incidentally, is the type of cosmos implied by the particle-physics theories that produce a value for the constant 10^{120} times larger than the observed value.) Careful calculations show that much smaller changes in other key parameters describing the physics of the cosmos could also remove the possibility of life. The strong anthropic principle holds that the cosmos must exist as it does because, in some sense, the cosmos requires observers. This idea contains overtones of quantum theory, but it does not hold much appeal for cosmologists. The strong anthropic principle does help us to refine what we are saying if we accept the weak version, not that the cosmos is as it is in order for us to be here, but rather that it is as it is because we are here.

Subtle distinctions can be powerful. Let us see how the anthropic principle relates to the most marvelous of the new cosmological approaches to the creation of the universe: the theory of the multiverse.

MANY UNIVERSES?

Current theories that attempt to explain the origin of the universe draw on the principles of quantum mechanics. These include the famous uncertainty principle, which states that attempts to measure one quantity with increasing accuracy must inevitably increase the uncertainty in our knowledge of a complementary quantity. Thus, for example, if we seek to determine a particle's position with near-perfect accuracy, we lose information about the particle's momentum (mass times velocity). This is not simply an issue of poor measurement techniques: The uncertainty principle implies that the combination of position and momentum can never be known with complete accuracy, so that the very notion of determining it must yield to a basic level of uncertainty. Within the white dwarfs that we met in Chapter 8, the exclusion principle represents a manifestation of the uncertainty principle. A white dwarf's self-gravitation, which squeezes its matter into a relatively tiny volume, attempts to give its electrons specific positions. The uncertainty principle implies that as the squeezing fixes the electrons' positions more exactly, their momenta become more uncertain, in order for the product of the uncertainties in the positions and momenta to exceed the minimum value set by the uncertainty principle. Hence the electrons cannot all have momenta extremely close to the specific value of zero. Instead, the uncertainty principle effectively endows the electrons with momenta, and therefore with velocities, which keep the electrons in continuous motion and thus prevent the white dwarf from collapsing to a single point.

To describe the universe fully, we need a theory that unites Einstein's general theory of relativity with the well-verified insights of quantum mechanics. Creating this combination has proven fiendishly difficult—not even an Einstein could achieve it, nor the finest theorists who have attacked the problem for decades. Some progress has apparently been made, however. The "superstring" theory of how elementary particles interact envisions the universe as containing ten dimensions, most of which, we may easily see, have been rolled up like a sock—"compactified," to use a technical term—so that they do not exist in the same way that the familiar dimensions do. Superstring theory will never play on the big screen, but its conclusions do not contradict what we know about elementary particles, and they offer testable predictions that can establish whether or not we should entrust our belief to the theory's explanatory powers. Nevertheless, for a while at least, we must confront the universe without a complete melding of quantum theory and general relativity to guide us. When we achieve this melding, we may glimpse the "grand unified theory" of which Einstein and his successors dreamed, a TOE, or theory of everything, capable of explaining all the forces in the universe. For now, we must each stand on our own two feet without a TOE, the lack of which does not—must not—prevent us from glimpsing a wonderful property of the cosmos: the ability to generate new universes.

Suppose we accept the concept that the universe (which we shall soon call "our universe") came into existence from a quantum fluctuation, a cosmic application of the uncertainty principle. Modern approaches to this problem, TOE-deprived though they may be, imply that before this occurred, nothing existed, quite literally (or mathematically)—not even time itself. But these same approaches imply in addition that the quantum fluctuation that created the universe need not be a unique event. At any instant, a quantum fluctuation could create a new universe, with its own big bang and subsequent evolution. We might imagine that the sudden creation of a universe would represent a serious problem of cosmic crowding. In fact, however, the mathematics behind the concept also implies that a new universe appears nowhere within the old! The newborn cosmos creates its own space and time, instead of taking space or running on another universe's time.

Holding in abeyance our natural, intuitive response that no such birth of a universe could occur (here practice helps, because the response will never disappear), we may admire the theory's connotations of universes busy being born while others are dying.

(Actually, a universe, once born, will never die, no matter what happens to the stars and galaxies within it.) What we call "the universe" could turn out to be just one among an infinite number, all born at different times and with different physical constants governing their development. So far as theory can tell, we have no chance of interacting with any cosmos but our own, a fact that justifies our continuing use of the word "cosmos" or "universe" to describe one among an infinity of universes or cosmoses. Nevertheless, their existence, even if demonstrable only at the level of cosmological theory, helps to set the anthropic principle in perspective and almost certainly to increase its appeal.

If an infinite number of universes exist, the fact that we find ourselves in one that is suitable for life appears to have a straightforward explanation: Life arises in universes that are fit for life. Likewise, we occupy a cosmos in which Ω_M and Ω_Λ are roughly equal because those are the cosmoses in which we may reasonably expect to find life. More precisely, we have seen that the universes with comparable values of Ω_M and Ω_Λ form a modest, but not a tiny, subset of those universes in which life can appear—the universes with physical laws that lead to the possibility of life plus an age (if those laws are similar to the ones in our own universe) measured in billions of years, rather than hundreds of millions or trillions.

The possibility of multiple universes leads to the conception and title of a "multiverse," the set of all universes. On the definitional front, the multiverse would replace what we have called "the universe" as the entity containing all that exists. The notion of a multiverse leads some theoreticians naturally toward the question, Does the multiverse itself evolve over time? In particular, does natural selection occur within the multiverse? Lee Smolin, a cosmologist at Pennsylvania State University, has suggested that some universes, thanks to the physical laws that govern them, may exhibit a greater tendency to produce new universes than others do. Then, just as natural selection rewards those organisms whose progeny survive in greater numbers, the multiverse would preferentially produce universes that tend to give birth to larger numbers of further universes. This assumes, of course, that the evolution of the multiverse follows Mendelian rather than Lamarckian genetics— that the characteristics of a universe that make it better at producing new universes can be passed from mother cosmos to daughter cosmos. The Lamarckian alternative might take us to the enjoyable speculation that universes better at restructuring themselves could carry the palm in the cosmic reproduction game, so that if we want our universe's characteristics to propagate, we had better improve matters around the cosmos. We would not, after all, want those unknown, quite possibly hostile, inhabitants of another universe to dominate the multiverse simply because we lacked the initiative to improve our territory.

Quintessence:
A Changing Cosmological Constant?

Theoretical cosmologists, straining mightily to explain a nonzero cosmological constant, are now investigating a class of theories in which the cosmological constant changes with time. This behavior, reminiscent of the cosmological constant that existed during the inflationary era, holds promise, if only because a changing cosmological constant has a better chance of ending up close to the actual value than a truly constant value does. As we discussed in Chapter 6, these cosmologists have resurrected the term "quintessence" to describe a cosmological constant that evolves with time, rather than remaining constant. Just as soon as the theorists produce a reasonable explanation of the value implied by the supernova observations, authors will rush their books back into print with revised editions incorporating those explanations.

The Greatest Cosmological Riddle

Speculation over Mendelian versus Lamarckian inheritance among universes probably represents a sufficient return of investment from the multiverse concept. Let us pause to notice how neatly this idea draws attention from the greatest of all cosmological mysteries: existence itself. Why is there something rather than nothing? Like the origin of life, which has proven a much greater riddle than life's evolutionary history, the origin of the cosmos poses a sterner problem than deciphering its past to predict its future. The multiverse concept takes this problem and casts it away as far as possible, into infinite recesses of time and across cosmic boundaries. All may agree, however, that the problem remains: How did the multiverse itself begin? If we someday achieve a

TOE, which might be renamed a TOEM (theory of everything in the multiverse), we may find an answer to this question, perhaps along the same quantum-mechanical, uncertainty-principle lines outlined above to explain our own poor universe. Yet logic suggests that whatever explanation may be offered, further explanations of the explanation will prove necessary.

Must we, then, abandon hope of finding a scientific explanation of the cosmos or cosmoses we have come to love so well? Must we bow to William Blake's strictures against those whom he saw as opposing revealed truth:

> *Mock on, mock on, Voltaire, Rousseau;*
> *Mock on, 'tis all in vain!*
> *You throw the sand against the wind,*
> *And the wind blows it back again.*

No good answer exists to satisfy all questioners. Both science and religion share the common ground of believing that the cosmic truth is out there, independent of anything humans can do or believe. Both seek to answer fundamental questions, such as the origin of everything. Both fall short in different ways, science because it must always test its accepted notions while seeking new ones to explain still more, religion because an appeal to higher authority cannot satisfy human longings to know the full story.

If we want more, we must do more. The quest to understand the cosmos has far more than spiritual significance to cosmologists and astronomers. An active life awaits, replete with instruments only dreamed of a decade ago. Let us look to the immediate human prospect as it stands ready to reveal the future of the cosmos.

CHAPTER FIFTEEN

PROSPECTS FOR
RESOLVING THE
COSMIC MYSTERIES

THE GOLDEN AGE OF COSMOLOGY, which dawned with Einstein's theory and Hubble's observations, entered its most glorious phase—so far!—with the use of Type Ia supernovae as accurate probes of the universe. By revealing the past, supernova explosions in distant galaxies have allowed cosmologists to predict a cosmic runaway future, subject to corrections arising from possible errors and misjudgments in interpreting the observational data.

Supernova observations, however, have provided only one significant portion of our new knowledge. Other methods for probing the past offer still-greater possibilities for the near-term revelation of cosmic truths, if only because they have not quite reached the developmental stage achieved by the supernova observers. We may confidently predict that within the next few years, new astronomical observations should resolve many of the most pressing issues of cosmology, as new satellites and other instruments extend our view deeper into space and time, while simultaneously embracing wider ranges of the electromagnetic spectrum.

Tradition and experience both support this prediction. During the 1940s and 1950s, astronomers believed that they would soon make accurate determinations of the Hubble constant and the total density of matter, which almost all believed resided in stars. In the 1960s and 1970s, astronomers confirmed the existence of dark matter in various guises, raising the stakes and the difficulty involved in finding the average density of matter in the universe. During the early 1980s, the inflationary model appeared, sounding a theoretical call for a precise value of the density, though even the most generous estimates of the dark matter left its density short of what theory demanded. In the early 1990s, a controversy over the value of the Hubble constant made some cosmologists doubt the full validity of the big-bang model, until improved observations and new calculations of stellar aging restored a proper balance between theory and observation. And as the 1990s reached their end, supernova observations implied a nonzero cosmological constant, capable of accelerating the universal expansion and providing an omega sufficient to raise the total omega to 1 and thus to validate the inflationary model. Though it now seems rather unlikely, we may yet find that systematic effects, such as the gray dust we imagined in Chapter 10, have fooled those who interpret supernova observations, so that a zero cosmological constant will return to its former most favored position.

In view of this history, why should the reader believe those who announce that the combination of new satellites to observe the cosmic background radiation, new studies of gravitational lensing, and new observations of supernovae at distances even greater than those so far studied will resolve cosmology's current conflicts and apparent contradictions? Because we have entered a new millennium; because we must remain optimistic about our abilities to perceive and to understand the cosmos; because "resolve" is a marvelously elastic word. If we choose to be brutally frank, we must admit that the flood of new data that certainly will emerge during the next decade may or may not produce general agreement about the crucial parameters that govern the evolution of the universe. If such agreement arises, new theories of the cosmos may yet lead to new debates over our place in the multiverse. This is good—good for the soul, good for the nation and the world.

Future Observations of
High-Redshift Supernovae

Our examination of the rewards of observing distant supernovae has demonstrated that the farther we look into space, the more we can tell about the competing influences of the density of matter and the cosmological constant. The news of a nonzero cosmological constant arrived from supernovae with redshifts between 0.3 and 0.7, which take us back in time to epochs when the universe had an age between 70 and 45 percent of its present value. This suffices to reveal times when the universe was significantly younger than it is now, and thus to provide significant evidence of a cosmic acceleration, caused by a cosmological constant. Even better, then, would be to observe still more distant supernovae, those with redshifts slightly greater than 1, which will take us back to a time when the universe had only about one-third of its present age. If astronomers can make accurate observations of the spectra and apparent brightnesses of these supernovae, they can determine whether or not the observed deviations in the Hubble diagram—away from the line describing a universe with no cosmological constant—do arise from the existence of a cosmological constant, or whether, instead, they have been misled by systematic effects.

What will it take to make these observations of fainter, more distant supernovae? More time on the world's great telescopes, especially the giant Kecks in Hawaii and the Hubble Space Telescope, which was due for a refurbishing mission at the end of 1999 but which will soon thereafter return to astronomical observations. Thanks to the success of supernova observations, obtaining this cherished time has become an easier task than before. Both groups of supernova experts—the experts of the Supernova Cosmology Project and those of the High-Z Supernova Search Team—have been assigned significant amounts of time on large telescopes, including not only the Kecks, but also the Canada-France-Hawaii Telescope on Mauna Kea and the Very Large Telescope in Chile, which will soon include four 8-meter telescopes. Time on the HST, which was scheduled to acquire a new infrared camera by the end of 1999, will also come to the supernova observers. Thus, although they understandably issue no guarantees, by the time that the new millennium really begins in 2001, the two supernova groups should have in hand significant amounts of observational data for supernovae at redshifts close to and slightly greater than 1. Analysis of the data, which may require a year of effort, will then yield a more definitive answer about the cosmological constant.

As described in Chapter 10, if the data from supernovae with redshifts greater than 1 reveal a Hubble diagram whose redshift-distance relationship shows an increasing deviation from the line that was anticipated before the revolution began, then either gray dust or systematic differences between the white dwarfs that produce distant and nearby Type Ia supernovae will offer the likeliest explanation of the great news from 1998. If, on the other hand, the cosmic track through the Hubble diagram reverts to the zero-cosmological-constant line at the highest redshifts, that will verify the assertion that we live in a universe with a nonzero cosmological constant and that we are looking sufficiently far back in time to see epochs when gravity's effects dominated those from the hidden energy in empty space.

Searching for Dark Matter

The truly big-picture cosmologists care most of all about the total density of matter in the universe, which determines the value of Ω_M, and comparatively less about the details of what's the matter. Nevertheless, like particle physicists who would dearly love to see their theories tested, cosmologists would welcome the detection of any of the forms that the unknown dark matter has been alleged to assume. They (and we) can live with the present situation, in which most of the matter in the universe consists of completely unknown types, but everyone would feel a good deal better if those types could be identified, or at least winnowed down by category.

Particle-physics theorists have proposed at least three excellent candidate types, each of which could turn out to provide most or all of the dark matter. (The obvious corollary, that quite possibly none of these particle types has more than a theoretical existence, must be stated if not celebrated.) One of these, the axion, would have a tiny mass, not much more than one-trillionth of an electron's mass. If the dark matter consists of axions, their numbers must therefore rise past the merely staggering. Another type of particle, the neutralino, would be more massive than a proton or a neutron; and the third, neutrinos with mass, would each have a mass about 1/17,000 of an electron's mass. Neutrinos, which come in three different types, belong to the category of actually existing particles, but the notion that any type of neutrino has a mass like that mentioned above remains speculative.

Axions and neutralinos appear in speculative extensions of the well-grounded and well-accepted theories of particle physics. These theories also call for the existence of neutrinos with masses greater than zero, possibly sufficiently large for neutrinos to constitute the bulk of the dark matter. All three types of particles, if they exist, would have been produced in enormous numbers soon after the big bang, and they would have dominated the mass budget of the cosmos ever since. If experiments could verify the existence of any one of these types of particles, the champagne would flow at many a laboratory and by many a patriot hearth. The attempts to find axions or neutralinos, and to determine the masses of neutrinos, have become more serious during the past few years. Like the search for extraterrestrial intelligence, the fact that particle physicists have yet to achieve success on the axion and neutralino front can be used by skeptics to "prove" that the hunt amounts to nothing but vain hope.

In fact, both searches are effectively in their infancies. If dark matter exists in particles of unknown form that fill the universe, all we need do is catch some of them, just as we might capture some of the conversations between other civilizations. Both attempts at capture are mightily hindered by the interference from Earthly processes and by the other forms of particles and radiation that stream onto the Earth. To find the dark-matter particles, physicists must build sensitive detectors and, to hunt for certain types of particles, place these detectors deep underground to avoid the interfering effects of fast-moving cosmic-ray particles.

If the dark matter turns out to be either axions or the neutrinos described above, cosmologists will have a problem, since they have already "proven" that most of the dark matter must be "cold" rather than "hot"—that at the time of decoupling, the majority of the dark-matter particles had speeds much less than the speed of light. The proof resides in computer models of how structures formed: The hot-dark-matter models cannot duplicate the cosmos we see today. In order for most of the dark matter to be cold, the mass per dark-matter particle must exceed the masses of axions or neutrinos by a substantial amount. We may freely speculate, however, that if particle-physics experiments detect axions or neutralinos and show that they exist in numbers sufficient to furnish the bulk of the dark matter, new computer models will be created, as they should be, to explain how the universe achieved its present state with hot rather than cold dark matter.

The absence of success in the search for the dark-matter particles emboldens us to pass over the details of experiments designed to detect them. If experimental data eventually eliminate axions, neutralinos, and neutrinos with significant mass as viable possibilities to comprise most of the dark matter, physicists will press onward to investigate the possible existence of other types of particles, some already hypothesized, others to spring from the brows of tomorrow's theorists. For now, cosmologists have the thrill of telling particle physicists about the "missing" matter in the universe, rather than listening to the physicists explain what different types of particles mean to cosmology. In actuality, the matter, far from missing, has been found from cosmologists' determinations of Ω_M. The buck has been passed to the particle experts, who continue to search for the particles that provide the bulk of the universe.

OTHER MEANS OF OBTAINING
the KEY COSMIC PARAMETERS:
THE SLOAN DIGITAL SKY SURVEY

As we have seen, attempts to detect Ω_M and Ω_Λ have implicated observations of gravitational lensing and of the complex structure in the universe, as well as the cosmic background radiation that we shall discuss in the following section. Astronomers will continue to study gravitational-lensing events with increased precision, and we may yet find that this tool yields the most definitive results for the crucial parameters. Presently, however, we may put this technique on hold, awaiting further results. Hardly the same can be said for observations of cosmic structure, which will take a great leap forward with one of the new millennium's grandest astronomy projects, the Sloan Digital Sky Survey. The SDSS, as astronomers refer to it, will map the visible universe far more deeply than any previous survey. By "deeply," it is meant that the SDSS will include galaxies at much greater distances than those recorded in other sky surveys.

Two-dimensional cosmological mapping is relatively easy; every photograph taken with a large telescope reveals galaxies and their position with respect to one another. Adding the third dimension, the depth of the universe and its objects, is by comparison awesomely difficult. If the astronomical efforts described throughout

This Hubble Space Telescope photograph shows a "deep field," revealed by an exposure lasting for dozens of hours, in the opposite direction as that shown in the photograph in Chapter 1. Like the first deep-field exposure, this one reveals galaxies older than 10 billion years, many of which formed only 1 or 2 billion years before their light left on its journey toward the Milky Way. (Photograph courtesy of the Space Telescope Science Institute.)

this book demonstrate one thing more than others, it is the difficulty of determining the distances to faraway galaxies. And yet—thanks to the efforts of Hubble and his successors—for galaxies closer than a few billion light-years, this task can be accomplished with relative ease.

Astronomers now know the value of the Hubble constant H within an accuracy of 10 percent, and they have created the Hubble diagram to show the relationship between galaxies' redshifts and distances. Once astronomers have made the painstaking observations that delineate the path that the redshift-distance relationship takes through the Hubble diagram, every measured redshift will correspond to a particular distance. For redshifts between 0.01 and 0.3, this redshift-distance relationship appears almost ironclad. Any inaccuracy in the distance derived from an object's redshift then arises almost entirely from our determination of the Hubble constant and should likewise not exceed 10 percent.

Previous deep-sky surveys have diligently obtained the spectrum of galaxy after galaxy, then analyzed those spectra to find the galaxy's redshift, and thus its distance, by the use of Hubble's law. Led by Margaret Geller and John Huchra at the Harvard-Smithsonian Center for Astrophysics, these multiyear efforts have yielded the redshifts of thousands of galaxies, located at distances out to hundreds of millions of light-years from the Milky Way.

The SDSS aims to map galaxies not by the thousands, but by the hundreds of thousands. To do so, the astronomers will not only secure many thousand spectra of individual galaxies, but will also employ a method to estimate galaxies' redshifts more rapidly than spectroscopic observations can. That method consists of measuring each galaxy's apparent brightness in five different colors. For relatively nearby galaxies, astronomers have already acquired a deep and thorough understanding of the brightness ratios that galaxies display in different colors. By measuring the ratios of brightness in these colors for distant galaxies, they can obtain an estimate, crude in comparison to an actual spectrum but sufficiently accurate for mapping purposes, of the galaxies' redshifts. Without this shortcut, the plan to make a three-dimensional map that includes a million galaxies and a hundred thousand quasars could never reach fruition.

Even so, the Sloan Digital Sky Survey, named after its chief funder, the Alfred P. Sloan Foundation, represents a mammoth undertaking, even though it will survey just over one-quarter of the total sky. The SDSS will perform this survey by measuring the colors of galaxies in nearly 2,000 regions on the sky, each of which covers an angular extent about thirty times greater than the full moon does. To accomplish the spectroscopic portion of their survey, the SDSS astronomers will prepare an aluminum plate for each region, drilled with 600 holes. The locations of these holes correspond to the positions of the brightest objects in that particular region on the sky. A fiber-optic cable will carry the light from each hole to CCD detectors that record the light from each object. On every clear night, the SDSS astronomers hope to observe six to nine of the regions, each of which yields more than thirty gigabytes (billions of bytes) of data. The task of storing and processing the data from all the observations staggers the weak mind: The SDSS will accumulate hundreds of terabytes (trillions of bytes—not yet a word in general usage) of data before completing its survey.

The fruit of this undertaking, by far the deepest map of the cosmos around us, will be subjected to intense comparison with the output of theoreticians' models. With any luck at all, the first decade of the twenty-first century should allow these comparisons to discriminate among the models, singling out one particular set for special attention for its link with reality. The observing program that was described in Chapter 12 will then have achieved the payday of which cosmologists now only dream.

Thirty Ways to Observe
the Cosmic Background Radiation

In Chapter 11, we saw that increasingly accurate observations of the cosmic background radiation offer the possibility of deciphering nearly the full set of parameters that describe the universe. This possibility arises for two reasons: The background radiation carries the complete record of conditions in the universe at the era of decoupling, and the details of how we see that radiation depend on the key parameters Ω_M and Ω_Λ.

To underscore the cosmological significance of accurate measurements of the cosmic background radiation, at least thirty separate experiments are now under way or in the planning and construction phases. These experiments involve arrays of radio telescopes at ground-based observatories, which can observe the longest-wavelength component of the background radiation; radio antennas at the South Pole and on a three-mile-high plateau in Chile, which can profit from the relative lack of atmospheric water vapor to extend these observations to somewhat shorter wavelengths; balloon-borne detectors, which rise above nearly all the water vapor to observe the radiation at the wavelengths that include its maximum energy output; and, most of all, the two future Earth-orbiting satellites that may sweep the board by providing so much information that the cosmos falls into place without further ado.

That last statement may have taken us a bit too far into Pollyanna physics, but every cosmologist eagerly anticipates, with good reason, the results to be sent to Earth by the satellites that will build on *COBE*'s results to map the details of the cosmic background radiation. As explained in Chapter 11, one of these satellites, NASA's *MAP* (an acronym, you may recall, for *Microwave Anisotropy Project*), will enter orbit, if all goes according to plan, at the end of the year 2000. The other, to be built by the European Space Agency (with some NASA contribution) is named *Planck*, which is not an acronym but a way to honor the great German physicist Max Planck, who opened the twentieth century with a concept of electromagnetic radiation that brought the quantum theory into existence.

The *Planck* satellite will study the cosmic background radiation with an accuracy of two parts per million and with an angular resolution twice as fine as *MAP*'s. These observations will extend our measurements of the intensity of the cosmic background radiation at different angular scales out to multipoles close to 1,000—that is, down to angular sizes as small as six minutes of arc.

In addition, *Planck* will have a capability much better than *MAP*'s for observing the cosmic background radiation in different polarizations. "Polarization" describes the direction of oscillation of the electric and magnetic fields embodied in beams of electromagnetic radiation. Most of the light and other forms of electromagnetic radiation produced by natural processes have no particular polarization: All directions of oscillation appear in equal amounts. Certain physical processes, however, produce radiation in which some directions preferentially appear, so that the radiation is said to be polarized. For example, when light or another type of electromagnetic radiation bounces off an electron, the directions in which the radiation scatters are not all equally likely, and the radiation observed after the scattering occurs is polarized—a clue to the fact that this scattering has occurred. In general, the amount and direction of any observed polarization bring additional information to astronomers and can provide valuable clues that reveal the processes that produced the radiation. Cosmological theories explaining how the universe produced the cosmic background radiation imply that we should observe a different amount of polarization at each different angular scale. *Planck's* ability to measure these polarizations will provide an impressively accurate check on different models of the early cosmos.

GRAVITY WAVES:
THE FINAL FRONTIER (FOR A WHILE)

If the *Planck* satellite works as well as planned, it may even be able to detect the effects of gravity waves in the very early universe. Gravity waves are ripples in space itself, completely different from electromagnetic radiation. Predicted by Einstein as a direct implication of his general theory of relativity, gravity waves have never been detected, though astronomers have observed their effects in the changing orbits of close binary stars as their motions produce gravity waves. The new millennium will see the start of operations of the Laser Interferometric Gravity-wave Observatory (LIGO), whose two identical installations, located in Louisiana and Washington State, each create a vacuum in four-kilometer-long tubes, set at right angles. Laser beams reflected back and forth through these tubes can reveal incredibly small vibrations of the masses suspended at the ends of the tubes to record the passage of gravity waves. The need for two installations arises from the fact that local disturbances ranging from a passing truck to a tiny earthquake can confuse a single detector.

LIGO may make the first direct detection of gravity waves, bringing Einstein's general theory of relativity down to Earth as it finds the ripples in space that arise when supernovae explode or double stars coalesce into a single object. Caught in the web of Earth, however, LIGO cannot detect the gravity waves that must have roiled the universe soon after the big bang, as space shook with the ripples of creation. For that we may use *Planck,* which will look for the effects of gravity waves on the background radiation, or we may use *LISA,* the *Laser Interferometer Space Antenna,* an ambitious NASA satellite that will create a LIGO in space, far from the interfering vibrations that our own planet produces.

Envoi

Humans come and go, but cosmology endures. "Roll on, thou deep and dark blue ocean, roll!" wrote Lord Byron, and the ocean did so. Likewise, the cosmos will proceed on its majestic evolution without regard to our knowledge of its past, present, and future. If we celebrate one thing more than another in the opportunities that life has given us, we might well choose the notion that our modest species, struggling as always for reproductive success, has taken the time to look into the cosmos and to understand—not completely, but more with each passing decade—the messages that arrive from distant galaxies, seen as they were billions of years in the past. With more such efforts, we may yet justify Oscar Wilde's noble observation that we are all in the gutter, but some of us are looking at the stars.

FURTHER READING

Adams, Fred, and Laughlin, Greg. *The Five Ages of the Universe: Inside the Physics of Eternity.* New York: Free Press, 1999.

Barrow, John. *Theories of Everything.* Oxford: Clarendon Press, 1991.

Chown, Marcus. *Afterglow of Creation: From the Cosmic Fireball to the Discovery of Ripples.* Sausalito, CA: University Science Books, 1996.

Christianson, Gale. *Edwin Hubble: Mariner of the Nebulae.* New York: Farrar, Straus, and Giroux, 1995.

Genz, Henning. *Nothingness: The Science of Empty Space.* Reading, MA: Helix Books, 1999.

Goldsmith, Donald. *Einstein's Greatest Blunder? The Cosmological Constant and Other Fudge Factors in the Physics of the Universe.* Cambridge, MA: Harvard University Press, 1995.

Greene, Brian. *The Elegant Universe.* New York: W. W. Norton, 1999.

Guth, Alan. *The Inflationary Universe.* Reading, MA: Helix Books, 1997.

Harrison, Edward. *Cosmology: The Science of the Universe.* Cambridge: Cambridge University Press, 1991.

Pais, Abram. *Subtle Is the Lord: The Science and Life of Albert Einstein.* Oxford: Oxford University Press, 1989.

Rees, Martin. *Before the Beginning.* Reading, MA: Helix Books, 1997.

Silk, Joseph. *A Short History of the Universe.* New York: Scientific American Library, 1994.

Smolin, Lee. *The Life of the Cosmos.* Oxford: Oxford University Press, 1997.

Struve, Otto, and Velta Zebergs. *Astronomy of the 20th Century.* New York: Macmillan, 1962.

Thorne, Kip. *Black Holes and Time Warps: Einstein's Outrageous Legacy.* New York: W. W. Norton, 1995.

Weinberg, Steven. *The First Three Minutes.* New York: Basic Books, 1973.

Weinberg, Steven. *Dreams of a Final Theory.* New York: Vintage Books, 1987.

INDEX

Aguirre, Anthony, 156–159
Albrecht, Andreas, 54
Alfred P. Sloan Foundation, 218
Alvarez, Luis, 115
Alvarez, Walter, 115
Anglo-Australian Observatory, 121
Anthropic principle, 204–206
Astronomical Society of the Pacific, 132
Average density of matter. *See* Matter, average density of
Axion, 214

Baade, Walter, 98–99, 144
Background radiation. *See* Cosmic background radiation
Balance, of universe, 12n
Balding, of astronomers, 92
Balloon model, of expansion, 34–35
Baryonic dark matter. *See* Dark matter
Batch process for supernova search, 123–125, 124(fig)
Berkeley. *See* Lawrence Berkeley Laboratory
Berkeley group. *See* Supernova Cosmology Project
Bethe, Hans, 98
Big-bang model, of expansion, 2, 212
  big crunch, 43
  causal contact, 57
  confirmation of through *COBE* data, 166–168
  era of decoupling, 163–164, 168–173, 178–179

flatness problem, 50–51, 56–59
horizon problem, 56–59
inflationary model and, 53–55
locating big bang in time, 38–39, 42
nucleosynthesis, 72–75
Big crunch, 43
Brahe, Tycho, 134
Brightness. *See* Luminosity

Canada-France-Hawaii Telescope, 213
Carter, Brandon, 204
Center for European Nuclear Research (CERN), 115
Cepheid variable stars
  brightness-distance correlation, 21–23
  classical vs. globular-cluster type, 144–145, 147
  estimating distances to, 141–143
  parallax effect, 27–28, 27n
  reexamining Hubble's data, 39
Cerro Tololo Inter-American Observatory, 125–126
Chandrasekhar, Subrahmanyan, 106
Chandrasekhar mass limit, 106, 110
Christianson, Gale, 20n
COBE. See Cosmic Background Explorer
Constants. *See* Cosmological constant; Hubble constant
Contraction, of the universe
  contraction, expansion, and static universe, 1–2, 10–12, 12n
  cosmological constant, 2–6
  general theory of relativity, 8–9

Copernican principle, 142
Cosmic Background Explorer (satellite),
　166–168, 171–177, 172(fig),
　178(fig)
Cosmic background radiation, 191–192
　COBE data on, 166–168
　directional uniformity, 52–53
　fluctuations in, 171–177, 172(fig),
　　178(fig)
　history of study, 164–166
　last scattering surface and, 168–171
　observational techniques, 219–221
　post-decoupling era, 183
Cosmological constant, 47, 201–202
　acceleration produced by nonzero
　　value, 78
　alternatives to, 92–93
　as quintessence, 55–56, 71, 71n, 209
　birth and death of, 10–12
　cosmic background radiation and,
　　176–177
　curvature of space and, 51
　density of matter, 46n
　diverse model universes, 44–45
　effect on expanding universe (*See
　　under* Matter, average density of)
　Friedmann-Lemaître models of
　　expansion, 37
　gravitational lensing, 197–198
　high redshift supernova
　　observation, 212–214
　inflationary universe and, 56n
　nonzero vs. zero value, 2–6, 13,
　　176–177
　rejection of zero value, 86–87, 89,
　　136–138
　resurrection of, 12–13
　rise times of SN Ia's, 151–152
　significance of nonzero value, 94–95
　SN Ia data, 147–148
　static universe and, 12n
　systematic effects on, 161
　vs. gray dust and systematic
　　differences, 159–160
　See also Ω_Λ
Cosmological principle, 32–34
Couch, Warrick, 120–121
Crab Nebula, 97

Curtis, Heber, 19–21
Curvature, of space, 8, 35–36, 36n,
　45–46, 50–51, 58, 168–171

Dark matter, 66–74, 106
　composition of, 214–216
　galaxy composition, 63–67,
　　191–192
　hot vs. cold, 185–187
　Hubble constant, 66–67
　light and, 62
　nucleosynthesis of, 72–74
　speed of light and, 185–187
　visible matter and, 63–67, 181–182
Decoupling, era of, 49, 163–164
　last scattering surface, 168–171
　spectrum of density fluctuations,
　　179
　temperature at, 171–173
Deep fields, 217(fig)
Density, of matter
　average density of (*See* Matter,
　　average density of)
　of intergalactic dust, 158–159
　variations during galaxy formation,
　　183–184
De Sitter, Willem, 11, 36
De Sitter model, of expansion, 36
Deuterium, 74–76
Dipole, 173
Distance measurement, 27–28, 131–132
　Cepheid variable stars, 21–23,
　　141–143
　effect of gray dust on measurement,
　　155–158
　Hubble diagram, 25–27
　Milky Way, 22–23
　miscalculations of, 143–147
　parallax effect, 27(fig)
　recession velocity-distance
　　relationship, 31–32, 133(fig)
　redshift-distance relationship,
　　212–214, 218
　search for standard candles, 25–27
　supernovae luminosity, 79–83,
　　126–127
　use of spectroscopy, 28–30
　See also Hubble diagram

Dominion Astronomical Observatory, 119

Doppler effect, 29(fig), 30–31, 34n, 63, 80, 154, 164–165

Dust, absorption of light by, 152–160

Eclipse, solar, 193–194

Edwin Hubble: Mariner of the Nebulae (Christianson), 20n

Einstein, Albert, 2–4, 9(fig), 20n
    cosmological constant, 10–12, 92–93
    gravity waves, 221
    Hubble and, 37–38
    predicting gravitational lensing, 192–198
    See also Cosmological constant; Relativity

Einstein arc, 195, 196(fig)

Einstein ring, 195

Elliptical galaxies, 23

Energy, 1–2
    cosmological constant and, 10–12
    supernova production of, 101–103

Exclusion principle, 206
    neutron stars, 106
    white dwarfs, 104–105

Expansion, of the universe, 1–2, 12
    age of, 38–39
    contraction, expansion, and static universe, 10–11, 12n
    cosmological constant, 2–6
    cosmological constant and rate of expansion, 12–13
    deceleration of, 47
    de Sitter model, 36
    Friedmann-Lemaître models, 37
    infinite prospect of, 94–95
    potential causes of acceleration, 92
    See also Cosmological constant; Relativity, general theory of

Explosions
    computer modeling of, 100
    of degenerate matter, 107–111

Faranda, Chuck, 148

Fermi National Accelerator Laboratory, 115

Filippenko, Alex, 132–138, 148–150

Flat model, of expansion, 35

Flatness problem, 50–51, 56–59

Friedmann, Alexander, 12n, 36–37

Fusion, nuclear, 100–101
    big bang nucleosynthesis, 75
    in white dwarfs, 108–109
    role in supernova production, 101–102

Galaxies, 5(fig), 153
    composition of, 63–67, 182–183, 191–192
    elliptical, 23
    formation and growth, 67–68, 169, 183–185
    galaxy clusters, 187–188
    gravitational deflection of starlight, 195–198, 196(fig)
    M81, 19(fig)
    redshift and recession velocity, 31–32, 41–42
    SDSS mapping project, 216–219
    types of, 22–23
    See also Distance measurement; Supernovae

Gamow, George, 12

Garnavich, Peter, 129

Geller, Margaret, 218

Globular clusters, 16

Goldhaber, Gerson, 140

Grand unified theory, 207

Gravitational lens arc, 195

Gravity, 216
    cosmological constant, 2–6
    dark matter, 63–67
    effect on expansion through time, 202–204
    effects over distance, 182–183
    galaxy formation and, 68
    gravitational lensing, 192–198, 193(fig), 196(fig)
    gravity waves, 221
    role in supernova explosions, 100–101
    types of, 68–71
    See also Relativity, general theory of

Gravity waves, 221

Great Debate, 19–21

Guth, Alan, 54

H. *See* Hubble constant
Hamuy, Mario, 136
Hansen, Leif, 119
Harvard Group. *See* High–Z
    Supernova Search Team
Harvard-Smithsonian Center for
    Astrophysics, 218
Harvard University, 127–128
    See also High–Z Supernova Search
    Team
Heisenberg uncertainty principle. *See*
    Uncertainty principle
High-Z Supernova Search Team,
    127–129, 136–138, 146, 213
Hipparcos (satellite), 27n
Homogeneity. *See* Smoothness
Horizon problem, 50, 52–53, 56–59, 170
HST. *See* Hubble Space Telescope
Hubble, Edwin, 15–19, 17(fig), 20n,
    25–27
    accomplishments of, 23
    de Sitter effect and, 36
    discovery of Cepheid variable stars,
    21–23
    first Hubble diagram, 31–32
    measuring distances to galaxies,
    27–28
    spectroscopy data, 30–31
Hubble constant (H), 32, 218
    age of the universe and, 77
    dark matter and, 66–67
    determining a value for, 39–40
    relation to density of matter, 46
    variation through time, 42–45, 46n,
    54–55
Hubble diagram, 84(fig)
    Einstein's acceptance of, 37–38
    high-redshift supernovae, 213–214
    history and future of the universe,
    43–45
    light-curves and luminosities,
    133(fig)
    modern version of, 44(fig)
    near vs. distant Type I supernovae,
    160
    original version, 31–32, 33(fig)

redshift-distance relationship,
    83–87, 218
    SCP results, 137(fig)
    special theory of relativity and,
    40–42
    supernovae as standard candles,
    79–83
Hubble's Law, 32–34, 38–39
Hubble Space Telescope (HST), 5(fig),
    124–125
Huchra, John, 218
Humason, Milton, 16, 38
Hydrodynamics, 100
Hydrogen–2. *See* Deuterium

Inflationary model, of the universe
    cosmological constant and, 53–56
    development of, 50
    horizon and flatness problems,
    56–59
    last scattering surface, 168–171
    open-inflation model, 59–60
    prediction of omega values, 89
    smoothness and, 189
International Ultraviolet Explorer
    (satellite), 128
Internet, 121, 129
Isaac Newton Telescope, 121

Jørgensen, Henning, 119

Keck telescopes, 135, 213
Kerrigan problem, 95–96, 203
Kirshner, Robert, 126–132, 146–147

Lagrange, Joseph-Louis, 173
Laser Interferometer Space Antenna
    (satellite), 221
Laser Interferometric Gravity-wave
    Observatory (LIGO), 221
Lawrence Berkeley Laboratory
    (formerly Lawrence Radiation
    Laboratory), 114–117, 120–122
Lawrence Livermore National
    Laboratory, 100
Lawrence Radiation Laboratory. *See*
    Lawrence Berkeley Laboratory
Leavitt, Henrietta Swan, 21, 142

Leibundgut, Bruno, 129
Lemaître, Georges, 36–37
Leuschner Observatory, 116, 118
Li, Weidong, 149
Lick Observatory, 19–21
Life, intelligent, 204–206, 208
Light
absorption by dust, 152–160
dark matter and, 62
Doppler effect, 30–31
gravitational lensing, 192–198,
193(fig), 196(fig)
Light, speed of
hot vs. cold dark matter, 185–187
inflationary era and, 55, 57–58
See also Relativity
Light curves, 131–132
LIGO. See Laser Interferometric
Gravity–wave Observatory
Linde, Andre, 54
LISA. See Laser Interferometer Space
Antenna
Los Alamos National Laboratory,
100
Lowell Observatory, 30–31
l-poles, 174, 177–178
Luminosity, of Cepheid variable stars,
25–28, 142, 144–145, 147
Luminosity, of supernovae, 97–98,
151–152
correlation to fade rate, 125–126,
131–132
determining redshift, 218
effect of dust on distribution of,
158–159
rise times to maximum luminosity,
148–150
supernovae as standards candles,
79–83
See also Light curve

MAP. See Microwave Anisotropy Project
Mapping, 216–219
Mass extinctions, 115–116
Mather, John, 166
Matter, 3
categories of, 70–71
degenerate, 104–106

effect on contraction and expansion,
2–6
gray dust, 155–160
inflationary universe, 212
See also Dark matter
Matter, average density of (Ω_M), 45–47,
46n, 50, 93–94
age of universe and, 78
cosmic background radiation,
172(fig), 176–177
dark matter, 66–67, 75–76
density as function of expansion,
202–204
determining through gravitational
lensing, 196–198
diverse model universes, 85–88,
88(fig), 186–187
flatness problem, 50–51
fluctuations in density, 188–190
future of the universe and,
113–114
galaxy formation and, 68
gravitational forces and, 182–183
inflationary universe and, 50, 58,
59–60, 89–92, 90(fig), 168–171
Kerrigan problem, 95–96
last scattering surface, 170–171
models vs. real universe, 191
observational values, 199(fig)
rate of expansion and, 202
SCP results, 139(fig)
M81 galaxy, 19
Microlensing, 195
Microwave Anisotropy Project (MAP),
177–179, 178(fig), 220
Milky Way, 16–19
distance to center, 22–23
Great Debate, 19–21
radiation from, 166–168
visible vs. dark matter, 182
Moon, as obstacle to observation,
122–123
Mount Stromlo Observatory, 129
Mount Wilson Observatory, 15–19, 22
Muller, Richard, 115–117
Multipole analysis, of background
radiation, 172(fig), 173–179
Multiverse theory, 206–210

National Science Foundation (NSF), 120
Nebulae. *See* Galaxies
Negative curvature model, of expansion, 35–36
Nemesis (solar companion star), 116
Neutralino, 214–216
Neutrino, 69–70, 214–216
Neutron stars, 99–103, 106
Newton, Isaac, 64
Nørgaard-Nielsen, Hans-Ulrik, 119
NSF. *See* National Science Foundation
Nucleosynthesis, 72–75

Observatories and telescopes
    Anglo-Australian Observatory, 121
    Canada-France-Hawaii Telescope, 213
    Cerro Tololo Inter-American Observatory, 125–126
    dark runs for supernova search, 122–123
    Dominion Astronomical Observatory, 119
    Hubble Space Telescope, 5(fig), 124–125, 217(fig)
    Isaac Newton Telescope, 121
    Keck telescopes, 135, 213
    Laser Interferometric Gravity-wave Observatory (LIGO), 221
    Leuschner Observatory, 116, 118
    Lick Observatory, 19–21
    Lowell Observatory, 30–31
    Mount Stromlo Observatory, 129
    Mount Wilson Observatory, 15–19, 22
    supernova observations, 213
    Very Large Telescope, 213
Octopole, 174
Ω_Λ, 45–47, 93–94, 160
    age of universe and, 78
    cosmic background radiation, 172(fig), 176–177
    determining through gravitational lensing, 196–198
    diverse model universes, 85–88, 88(fig)
    flatness problem, 50–51

future of the universe and, 113–114
    gravitational forces and, 182–183
    inflationary universe and, 58–60, 89–92, 90(fig), 168–171
    Kerrigan problem, 95–96
    last scattering surface, 170–171
    models vs. real universe, 191
    observational values, 199(fig)
    rate of expansion and, 202
    SCP results, 139(fig)
Ω_M. *See* Matter, average density of

Palomar Mountain, 134
Parallax effect, 26–27, 27(fig), 27n
Particle physics, 115, 189–190
    dark matter, 62–63
    inflationary era, 55–56
    value of cosmological constant, 201–202
Particles, subatomic
    as components of dark matter, 69–71, 214–216
    exclusion principle, 104–105
Payne, Cecilia, 29–30, 127
Pennypacker, Carl, 118
Perlmutter, Saul, 114–122, 126–127, 132, 150–151, 161
    See also Supernova Cosmology Project
Phillips, Mark, 11, 125–126, 136
Planck, Max, 220
Planck (satellite), 179, 220–221
Polarization, of electromagnetic radiation, 220–221
Proper motion, of stars, 143
Press, William, 131
Prussian Academy, 7

Quadrupole, 173
Quantum theory, 189, 206–209
Quintessence, 54–55

Radiation, background. *See* Cosmic background radiation
Recession velocity, 40–42, 140
    inflationary universe and, 49

recession velocity-distance
relationship, 31–32
relativity and, 140n
See also Hubble diagram
Reddening, of light, 153–154
Redshift, 79–83, 161
affect of reddening on, 153
converting to velocity, 40–42
determining age of universe, 165
high- vs. low-redshift supernovae,
147–152, 152, 188
new Hubble diagram, 83–87
recession velocity and, 140
redshift-distance relationship,
212–214, 218
See also Hubble diagram
Relativity, general theory of, 7–11
gravitational deflection of light,
191–198, 193(fig)
recession velocity and, 140n
Relativity, special theory of, 138–140
converting redshift to recession
velocity, 41n
inflationary universe and, 57–58
recession velocity and, 140n
Richardson, Harvey, 119
Riess, Adam, 129–132, 136, 148–151,
161
Rise time, of supernova luminosities,
148–151
Rubin, Vera, 65
Russell, Henry Norris, 18

Sargent, Wallace, 134
Scattering surface, 168
Schmidt, Brian, 127, 129, 136,
148–149
SCP. *See* Supernova Cosmology
Project
SDSS. *See* Sloan Digital Sky Survey
Shapiro, Irwin, 130
Shapley, Harlow, 16–23, 17(fig)
SLAC. *See* Stanford Linear Accelerator
Center
Slipher, Vesto, 30–31, 40
Sloan Digital Sky Survey (SDSS),
216–219
Smolin, Lee, 208

Smoothness, of cosmic background,
52–53, 167–177
SN Ia. *See* Supernovae, Type Ia
Space, curvature of, 8, 35–36, 36n,
45–46, 50–51, 58, 168–171
Spectroscopy, 28–30
Spiral galaxies, 18–23, 19(fig)
Spiral nebulae. *See* Spiral galaxies
Standard candles, 25–27, 125
SN Ia's as, 125–126, 129–132
supernovae as, 79–83
systematic errors in method,
145–147
Type Ia vs. Type II supernovae,
128–129
Stanford Linear Accelerator Center
(SLAC), 115
Static universe, 2–6
Steinhardt, Paul, 54, 60
Supernova Cosmology Project (SCP),
114–123, 126–127, 136–138,
139(fig), 146, 213
batch process supernova search,
123–125, 124(fig)
Supernovae, 2–5, 81(fig), 97–99
as standard candles, 79–83,
128–129
batch process for searches, 123–125,
124(fig)
classification of, 80
dark runs for observation, 122–123
effect of dust on observation of,
153–160
future observational challenges,
160–162, 212–214
high-redshift vs. low-redshift,
147–152
Hubble diagram, 83–87, 84(fig)
in binary systems, 109–111
light curve adjustments, 140
light curves of SN Ia's, 131–132
luminosity of, 125–126
near vs. distant, 159–160
nuclear fusion and, 101–102
revealing density parameters,
89–92
rise time of, 148–149
robotic search for, 116–125

systematic errors in distance
    calculations, 145–147
Type I, 106–107
Type Ia, 107–111, 118–120, 125–126,
    133(fig), 137(fig), 151–152, 161,
    176
Type Ia vs. Type II as standard
    candles, 128–129
Type Ib, 134–135
Type II, 99–104, 128–129
use in measuring cosmological
    constant, 3–4
See also High–Z Supernova Search
    Team; Supernova Cosmology
    Project
Superstring theory, 207
Systematic error, 151

Telescopes. *See* Observatories and
    telescopes
Temperature, at decoupling, 171–173,
    172(fig)
Time, slowing of, 138–140, 140n
Triangulation method of
    measurement, 26–27
Trumpler Prize, 132
Turner, Michael, 75, 95

Uncertainty principle, 206
Universe
    age of, 38–39, 42, 77–78, 87–89, 147n,
        165
    balance of, 12n
    computer models of formation,
        185–187

curvature of (*See* Curvature)
five constituents of, 70–71
future of, 113–114
Mendelian vs. Lamarckian
    inheritance, 208–210
multiverse theory, 206–210
Universe, models of, 85–88, 88(fig),
    137(fig), 160, 178(fig)
    big-bang model (*See* Big-bang
        model)
    computer models of formation,
        185–187
    cosmological constant and,
        44–45
    expansion models, 34–36
    fluctuation of cosmic background
        radiation, 172(fig)
    Friedmann-Lemaître models, 37
    inflationary model, 50, 53–60, 89,
        168–170, 189
    static universe, 1–6, 10–12, 12n

Van Maanen, Adrian, 20
Velocity. *See* Distance measurement;
    Hubble diagram
Velocity-distance relationship. *See*
    Distance measurement; Hubble
    diagram
Very Large Telescope, 213
Virgo Cluster, 187–188

White dwarfs, 104–111, 151–152
World War I, 7–9

Zwicky, Fritz, 63, 98–99, 127